The Kidney of the CRITICALLY ILL PREGNANT WOMAN

Critical Care in Obstetrics

The Kidney of the CRITICALLY ILL PREGNANT WOMAN

Series editors

CARLOS MONTUFAR-RUEDA

Critical Care Obstetric Unit, Dr. Arnulfo Arias M Hospital Complex, Panama City, Panama

JORGE HIDALGO

Division of Critical care, Belize Healthcare Partners, Belize

JAVIER PEREZ-FERNANDEZ

Critical Care Services Baptist Hospital of Miami, Miami Beach, FL, United States

ELSEVIER

ACADEMIC PRESS

An imprint of Elsevier

Academic Press is an imprint of Elsevier
125 London Wall, London EC2Y 5AS, United Kingdom
525 B Street, Suite 1650, San Diego, CA 92101, United States
50 Hampshire Street, 5th Floor, Cambridge, MA 02139, United States

ISBN: 978-0-443-21473-8

For information on all Academic Press publications visit our website at https://www.elsevier.com/books-and-journals

Publisher: Stacy Masucci
Acquisitions Editor: Patricia Osborn
Editorial Project Manager: Susan Ikeda
Production Project Manager: Fahmida Sultana
Cover Designer: Mark Rogers

Typeset by TNQ Technologies

Contents

Contributors

Ahsan Ahmed
Critical Care Medicine, KPC Medical College and Hospital, Kolkata, West Bengal, India

Hythem Al-Sum
King Abdulaziz Medical City, Riyadh, Saudi Arabia

Maisah Albakri
King Abdulaziz Medical City, Riyadh, Saudi Arabia

Maria Jimena Alemán
School of Medicine, Universidad Francisco Marroquin, Ciudad de Guatemala, Guatemala

Jomana Abdullah Al Mashikas
Department of Obstetrics & Gynecology, Qatif Central Hospital, Qatif, Saudi Arabia

Mohammed Alabdrab Alnabi
Department of Critical Care Medicine, Dammam Medical Complex, Dammam, Saudi Arabia

Nehad Nabeel Al Sherawi
Al Fujairah Hospital, EHA, United Arab Emirates

Suhad Alsurori
King Abdulaziz Medical City, Riyadh, Saudi Arabia

Juan Carlos Barrientos
High Dependency Obstetric Unit, San Juan de Dios General Hospital, Ciudad de Guatemala, Guatemala

Payel Bose
Critical Care Medicine, Medica Superspecialty Hospital, Kolkata, West Bengal, India

Anirban Bose
Medica Institute of Critical Care, Medica Superspecialty Hospital, Kolkata, West Bengal, India

Ranajit Chatterjee
Swami Dayanand Hospital, New Delhi, Delhi, India

Priyanka H. Chhabra
Department of Anaesthesia, VMMC-Safdarjung Hospital, Delhi, India

Luisa Maria Cruz
School of Medicine, Universidad Francisco Marroquin, Ciudad de Guatemala, Guatemala

Saurabh Debnath
Critical Care Medicine, B P Poddar Hospital and Research Ltd., Kolkata, West Bengal, India

Vera Álava Fabricio
Manta General Hospital – IESS, Critical Care Ecuatorian Society (SECI), Laica Eloy Alfaro University of Manabí, Manabí, Ecuador

Lalit Gupta
Anaesthesia and Intensive Care, Maulana Azad Medical College, New Delhi, Delhi, India

Ahsina Jahan Lopa
Department of ICU and Emergency, Brahmanbaria ICU Specialized Hospital, Dhaka, Bangladesh

Moreira Macías Jéssica
Ecuatorian Pichincha Colposcopy Society, Manta General Hospital – IESS, Manabí, Ecuador

Maria Cristina Jiménez Bazzano
Department of Internal Medicine and Clinical Nutrition, Faculty of Medical Sciences, National University of Asunción, Clinicas Hospital, Asunción, Paraguay

Kamal Lashkari
Department of Critical Care, Bombay Hospital, Indore, Madhya Pradesh, India

Rodríguez María Virginia
Critical Care Ecuatorian Society (SECI), Rafael Rodríguez Zambrano Hospital – MSP, Laica Eloy Alfaro University of Manabí, Manabí, Ecuador

Fatemeh Masjedi
Shiraz Nephro-Urology Research Center, Shiraz University of Medical Sciences, Shiraz, Iran

Md Mostafizur Rahman Akash
Shaheed Monsur Ali Medical College Hospital, Uttara, Dhaka, Bangladesh

Elvira Gimenez Rolón
Department of Internal Medicine-Nephrology, Faculty of Medical Sciences, National University of Asunción, Clinicas Hospital, Asunción, Paraguay

Jamshid Roozbeh
Shiraz Nephro-Urology Research Center, Shiraz University of Medical Sciences, Shiraz, Iran

Carlos Sánchez
Critical Care Medicine, King Salman Hospital, Riyadh, Saudi Arabia

Byron F. Sapon Perez
University of San Carlos of Guatemala, Guatemala City, Guatemala

Utpal Sarma
Critical Care, Sri Balaji Action Medical Institute, Paschim Vihar, New Delhi, India

Juan Ignacio Silesky-Jimenez
San Juan de Dios Hospital, Caja Costarricense del Seguro Social and CIMA Hospital, San José, Costa Rica

Vinay Singhal
Critical Care & Emergency Services, SPS Hospitals, Ludhiana, Punjab, India

Preface

The care of serious complications in pregnant or postpartum patients has become an increasingly necessary and demanding discipline. The formation of care units for pregnant women in critical condition is intended to support organs or systems in failure, which must have a medical team that understands the physiology of the pregnant patient, the pathophysiology of conditions that are unique in obstetrics, as well as the behavior of medical—surgical pathologies that may occur in patients during pregnancy and/or the postpartum period.

One of the organs that most transforms your work intensity is the kidney, since this organ ends up handling the large increase in volume. This ends in a significant increase in the glomerular filtration rate, and consequently, the kidney has an increase in its metabolic demands. This increase in kidney function, under normal conditions, achieves a great balance in the body distribution of fluids in pregnant women, but it makes this organ more susceptible to failure. In this way, we frequently find kidney lesions, both of prerenal and renal origin, in scenarios of preeclampsia, severe obstetric hemorrhages, complications of abortions, acute fatty liver of pregnancy, and others.

The aforementioned susceptibility of the kidney, due to its high metabolic demand, means that we must make earlier alerts and decisions, such as timely defining the risk criteria for damaging the kidney and intervening earlier in the decision of renal replacement therapy.

The literature is emphatic in determining that, in obstetric patients, permanent residual renal lesions are in a higher percentage than we suspect, generating a considerable number of patients in chronic renal failure in the long term.

Knowledge of renal medical conditions, as well as renal damage secondary to acute obstetric pathologies, is important to decide the correct intervention, as well as define multidisciplinary management to obtain the best results.

In this text, we wanted to collect our experiences and the best scientific evidence to show how to deal with the kidney damage that multiple acute medical and obstetric clinical conditions can cause, and in this way have the best tools to approach these patients, and obtain the best results.

We thank all the collaborators who developed the different chapters and topics of this book, to complete a work of help to health personnel who care for and deal with obstetric patients, specifically with kidney damage.

We also wish to thank our families, who have allowed us to use the time necessary to complete this work and have sacrificed their time.

CHAPTER 1

Pregnancy induced kidney physiological alterations

Luisa Maria Cruz and Maria Jimena Alemán
School of Medicine, Universidad Francisco Marroquin, Ciudad de Guatemala, Guatemala

Anatomic changes

Pregnancy has several physiological effects on the body, including changes to the kidneys. During pregnancy, the size of the kidneys increases due to an increase in volume and dilatation of the collecting system. The renal volume increases by about 30% and the overall dimensions of the kidney increase by about 1 cm [1]. This increase in size is thought to be due to a combination of circulating mediators and the dilatation persists postpartum.

The right renal pelvis is more frequently dilated than the left during pregnancy, and this is thought to be due to mechanical compression of the ureter by the uterus as well as the anatomical location of the right ureter [2,3]. The dextrorotation of the uterus and the presence of the sigmoid colon in the left pelvis may also contribute to this asymmetric dilatation [4]. Dilatation of the ureter is more common on the right-side during pregnancy and is thought to be primarily caused by mechanical compression by the enlarging uterus and the ovarian venous plexus [5].

The renal parenchymal volume increases likely due to increases in both vascular and interstitial fluid volume, rather than cellular hyperplasia or hypertrophy [6]. As a result, urinary stasis and hydronephrosis (with or without calyceal clubbing) are common physiological occurrences in pregnancy and do not usually reflect pathological obstruction. However, some women may experience "overdistention" syndrome in late gestation, which is characterized by marked hydronephrosis, abdominal pain, increases in serum creatinine, and in some cases, hypertension [6].This can be managed with ureteral stents, which can be removed postpartum without recurrence of symptoms. It is also possible for pregnant women to experience acute urinary retention and nontraumatic rupture of the ureter, but it is not clear if these are extreme instances of physiological ureteral dilatation or underlying pathology.

Overall, the changes to the kidneys during pregnancy are largely physiological in nature and resolve after delivery. While there may be some discomfort or complications associated with these changes, they do not usually indicate underlying pathology. It is important for pregnant women to be aware of these changes and to consult with their healthcare provider if they have any concerns.

The Kidney of the Critically Ill Pregnant Woman
ISBN 978-0-443-21473-8,
https://doi.org/10.1016/B978-0-443-21473-8.00018-5

Hemodynamic changes

Systemic changes

Normal pregnancy represents a great hemodynamic challenge because of the marked volume expansion and vasodilation that occurs with gestation [7]. Blood volume expands by an increase in both plasma volume, and to a lesser extent, an increase in red blood cell volume [7].

Maternal systemic vasodilation begins early in pregnancy, around the fifth week of gestation, caused by a decrease in systemic vascular resistance (SVR) of 35%–40% [8]. SVR decreases secondary to the creation of a low resistance uteroplacental circulation and vasodilatory factors such as prostacyclin and nitric oxide [9]. Other proposed mechanisms include a reduced vascular responsiveness to vasopressors such as angiotensin II, norepinephrine, and antidiuretic hormone, reduced aortic stiffness, and other hormonal agents (relaxin and progesterone) [10,11].

The most well-studied mechanism includes the renin angiotensin aldosterone system. Reduced SVR activates the renin–angiotensin–aldosterone system (RAAS) increasing renin, angiotensin II, and aldosterone. Additionally, renin is also released from extrarenal sources, such as the ovarium and decidua and angiotensin II production is increased because of high estrogen levels produced by the placenta [12]. Although with increased angiotensin II there is an expected vasoconstriction, in normal pregnancies, there exists a refractoriness to angiotensin II-mediated vasoconstriction [11]. Angiotensin I receptors, which cause vasoconstriction, are downregulated, while angiotensin II receptors, which cause vasodilation, are upregulated in pregnancy [13]. These changes in the RAAS system contribute to the vasodilatory state and volume expansion in pregnancy [14].

As maternal and fetal metabolic demands increase along the pregnancy, cardiac output also increases to meet such demands. Cardiac output increases by 50% secondary to changes in both stroke volume and heart rate [7,8]. Stroke volume is increased secondary to an increase in preload due to the expansion seen in blood volume and a reduction in afterload due to the decline in SVR. The contractility of the myocardium, which also affects stroke volume, may be slightly improved [15]. Maternal heart rate also increases during pregnancy [7].The net effect of the changes in the SVR and the cardiac output is a drop in systemic blood pressure.

Renal changes

The global hemodynamic changes are accompanied by an increase in renal perfusion and glomerular filtration [16]. Even before pregnancy, the glomerular filtration rate (GFR) increases during the second half of the menstrual cycle as anticipation of a possible pregnancy [17]. In the first half of the pregnancy, renal plasma flow (RPF) increases by up to 80% and the GFR increases by up to 40%–50% [16,18]. As the rise in RPF is greater than the rise in GFR, the filtration fraction is reduced. Late in gestation, RPF is reduced while GFR is maintained, resulting in an increased FF.

A reduction in the SVR is the primary driver behind an increase in renal perfusion which results in the increase in the GFR [19]. The exact mechanisms behind the changes in renal hemodynamics are yet to be fully elucidated; however, the specific increase in renal vasodilation is greatly influenced by the hormone relaxin. Relaxin is a 6-kD peptide hormone in the insulin family that originates during the menstrual cycle by the corpus luteum and during pregnancy by the placenta and decidua in response to human chorionic gonadotropin. Relaxin is a key renal vasodilator as it increases the production of endothelin and nitric oxide production in the renal circulation. It reduces renal afferent and efferent resistance causing a subsequent increase in RPF and GFR [20,21].

These changes cause a decrease in serum creatinine concentration, particularly in early pregnancy, to a normal range of 0.4—0.8 mg/dL. A serum creatinine of 0.8 mg/dL or higher, or a small rise in serum creatinine may indicate renal dysfunction in pregnant patients. To estimate GFR in pregnancy patients, a 24-h urine collection remains the standard of care as creatinine-based formulas that estimate GFR underestimate GFR in pregnancy [22].

Renal tubular changes

Osmoregulation and sodium

During normal pregnancy, serum osmolality decreases to a new set point of 270 mOsm/kg with a proportional average decrease in plasma sodium concentration of 4—5 mEq/L [9,23]. A decrease in osmolality of such magnitude would normally suppress the antidiuretic hormone response and cause a subsequent massive water diuresis; however, the osmotic threshold for an antidiuretic hormone response and thirst sensation are also lowered during pregnancy, suggesting a "reset of the osmostat" during gestation [1]. These adaptations result in water retention with an accumulation of 6—8L of water stored in the products of conception, maternal intracellular, and extracellular components [24]. β- human chorionic gonadotropin and relaxin are the main drivers behind the osmoregulatory changes in pregnancy [25,26].

The hyponatremia in pregnancy is the result of a complex interplay of natriuretic and antinatriuretic factors. The natriuretic effects of increased GFR, and elevated atrial natriuretic peptide and progesterone levels are balanced by the antinatriuretic effects of increased aldosterone, angiotensin II, and deoxycosterone levels [14]. Despite a total body sodium gain during pregnancy of around 1000mEq of sodium, there is a greater water gain resulting in hyponatremia [9]. A serum sodium concentration lower than 130mEq/L in pregnancy should be considered abnormal.

Potassium

Pregnancy is a unique physiological state that is associated with a number of changes in the body. One of these changes is an increase in the hormone aldosterone and mild alkalosis. Aldosterone is a hormone that plays a key role in regulating electrolyte balance, particularly in the kidneys, by promoting the reabsorption of sodium and the excretion of potassium. Mild alkalosis, on the other hand, refers to a slight increase in the pH of the blood due to a decrease in carbon dioxide.

Despite these changes, pregnancy is not associated with hypokalemia, which is a condition characterized by low levels of potassium in the blood. This is surprising, as increased aldosterone and mild alkalosis would typically be expected to lead to potassium loss and hypokalemia [27].

The reason for this lack of hypokalemia in pregnancy is thought to be due to the antimineralocorticoid effect of progesterone. Progesterone is a hormone that is produced in large amounts during pregnancy and plays a key role in maintaining the pregnancy. It has been shown to competitively inhibit aldosterone at the mineralocorticoid receptor, which prevents distal sodium reabsorption and potassium excretion [34]. This means that progesterone effectively counteracts the effects of aldosterone, preventing potassium loss and maintaining normal potassium levels.

Furthermore, progesterone has the capacity to inhibit potassium excretion independent of antimineralocorticoid effects. This means that progesterone can inhibit potassium excretion through mechanisms that are unrelated to its effects on aldosterone. This further helps to maintain normal potassium levels during pregnancy [14].

In conclusion, pregnancy is associated with an increase in aldosterone and mild alkalosis, which would typically be expected to lead to hypokalemia. However, this is not the case in pregnancy, due to the antimineralocorticoid effect of progesterone, which prevents potassium loss and maintains normal potassium levels. Additionally, progesterone also has the capacity to inhibit potassium excretion independent of antimineralocorticoid effects. This helps to further maintain normal potassium levels during pregnancy.

Calcium and phosphorus

During pregnancy, the placenta plays a vital role in supporting the development of the fetus. One important function of the placenta is the production of parathyroid-related peptides, which stimulate the production of vitamin D (1, 25) [28].This increase in vitamin D leads to an increase in maternal calcium gastrointestinal absorption, which is essential for fetal skeletal development.

However, while calcium absorption is increased, total calcium levels in the mother's body may drop due to a decrease in serum albumin. Maternal ionized calcium levels remain unchanged. Magnesium levels, both total and ionized, also drop during pregnancy [29].

To compensate for these changes, the mother's body increases urinary excretion of calcium, as well as magnesium, citrate, acidic glycoproteins, and nephrocalcin. These substances help prevent the formation of calcium oxalate stones, which can be a potential problem due to the calcium supersaturation that occurs during pregnancy.

Overall, the changes in calcium and magnesium levels during pregnancy are necessary to support the development of the fetus and to maintain the mother's health. The placenta plays a crucial role in these processes through its production of parathyroid-related peptides and the regulation of vitamin D levels.

Uric acid

During pregnancy, there are changes in the levels of serum uric acid in the body. Early in pregnancy, uric acid levels fall to between 2.0 and 3.0 mg/dL [27]. However, after 22–24 weeks of pregnancy, uric acid levels start to rise due to an increase in uric acid production as a result of trophoblast breakdown, ischemia, and the release of cytokines [30]. The renal tubular absorption of urate also increases during this time. By 37 weeks of gestation, serum uric acid levels can reach nonpregnant levels.

The mechanisms behind the changes in uric acid levels during pregnancy are not fully understood, but altered renal handling is thought to be the major contributing factor. This can include increased GFR and reduced proximal tubular reabsorption of urate, or a combination of both. Certain conditions, such as preeclampsia, which is characterized by plasma volume contraction, can lead to decreased uric acid clearance and increased circulating levels of uric acid. Uric acid has received significant attention in the obstetric literature for its role in the prediction, diagnosis, and pathophysiology of preeclampsia [16,31].

Glucose

Glycosuria, or the presence of glucose in the urine, is a common occurrence during pregnancy. This is due to increased glucose filtration by the kidneys coupled with decreased tubular reabsorption, which is thought to be a result of increased plasma volume during pregnancy. When the filtered load of glucose reaches the maximum resorptive capacity of the proximal tubule, there is physiologic glycosuria, although serum glucose concentrations typically remain normal [27].

Normally, glucose is freely filtered at the level of the glomerulus and almost completely reabsorbed by a sodium-coupled active transport process in the proximal tubule [1]. Some glucose is also absorbed in the collecting tubule. However, during pregnancy, the increased GFR and tubular flow rate may limit the ability of the proximal tubule to completely reabsorb glucose, resulting in physiologic glucosuria. This is a normal physiological occurrence during pregnancy and does not indicate any underlying pathology.

Acid base

The kidneys play an important role in maintaining acid-base homeostasis during pregnancy [1]. One of the primary changes that occurs in pregnancy is a relative respiratory alkalosis, which is caused by an increase in maternal minute ventilation. This reduces the maternal partial pressure of carbon dioxide (pCO_2) as measured on arterial blood gas analysis and serves as an important gradient for gas exchange between the mother and the fetus.

To compensate for the respiratory alkalosis, the kidney increases bicarbonate excretion, which leads to reduced serum bicarbonate levels [1]. This decrease in bicarbonate levels may limit the buffering capacity of the body during pregnancy.

Protein

Proteinuria, or the presence of protein in the urine, is a common occurrence during pregnancy. In normal pregnancy, the level of urinary protein excretion increases from approximately 100 mg/day in the nonpregnant state to 150–200 mg/day in the third trimester [32]. This increase may result in a positive dipstick result when a concentrated urine sample is examined. However, urinary protein excretion greater than 300 mg/day is considered abnormal and requires further evaluation [33].

The mechanisms behind the increase in urinary protein excretion during pregnancy are not fully understood but may include an increase in GFR, an increase in the pore size of the glomerular basement membrane, increased transport of protein across the glomerular filtration barrier via the nondiscriminatory shunt pathway, and reduced tubular reabsorption of filtered protein. Circulating antiangiogenic factors, which cause glomerular endothelial dysfunction and proteinuria in preeclampsia, also increase toward term, and may contribute to late gestational proteinuria even in the absence of preeclampsia.

The ratio of urinary protein to creatine has been found to rise more than expected based on the albumin-to-creatinine ratio, indicating that a nonalbumin protein may be responsible for the mild increase in urine protein excretion. This makes it less likely that the proteinuria is due to "glomerular leak" from hyperfiltration, as albumin would be the expected excreted protein in that case [14]. Therefore, the appearance of new proteinuria or worsening of proteinuria in a pregnant patient warrants clinical evaluation for possible underlying kidney disease or preeclampsia. If preeclampsia is ruled out, the same diagnostic evaluation should be carried out as for a nonpregnant individual.

Other substances

Water-soluble vitamins are essential for maintaining good health, and they are easily excreted by the kidneys. However, during pregnancy, the renal excretion of these vitamins increases, which may put pregnant women at risk of deficiency. This increased excretion may be due to changes in the renal tubules that lead to reduced reabsorption of these vitamins.

Nicotinic acid, also known as niacin, is a B-vitamin that is essential for the metabolism of carbohydrates, fats, and proteins. It is also necessary for maintaining healthy skin, nerves, and digestive system. During pregnancy, the increased renal excretion of nicotinic acid may put pregnant women at risk of deficiency, which can lead to pellagra, a disease characterized by skin lesions, diarrhea, and neurological symptoms.

Ascorbic acid, also known as vitamin C, is a powerful antioxidant that is essential for maintaining healthy skin, bones, and connective tissue. It also plays a vital role in the immune system and helps the body absorb iron. The increased renal excretion of ascorbic acid during pregnancy may put pregnant women at risk of deficiency, which can lead to scurvy, a disease characterized by fatigue, muscle weakness, and bleeding gums.

Folic acid, also known as vitamin B9, is essential for the formation of red blood cells and the proper development of the nervous system. During pregnancy, the increased renal excretion of folic acid may put pregnant women at risk of deficiency, which can lead to neural tube defects in the developing fetus.

Interestingly, plasma folate levels decrease in pregnancy, this could be due to the increased demand of the developing fetus for folic acid, which leads to a depletion of maternal stores. The increased urinary excretion of folic acid may be primarily caused by reduced tubular reabsorption rather than an increased filtered load.

Conclusion

In conclusion, pregnancy results in significant changes in the renal system, including volume expansion and vasodilation, alterations in systemic and renal hemodynamics, and increased RPF and GFR. These changes are necessary to accommodate the demands of the fetus and ensure the health and well-being of both the mother and the baby. Renal tubular function also responds to these changes by altering water and electrolyte handling, resulting in lower serum osmolality, mild hyponatremia, and increased glucose and protein in the urine.

References

[1] Gammill HS, Jeyabalan A. Acute kidney injury in pregnancy epidemiology of acute kidney injury in pregnancy.

[2] Faúndes A, Brícola-Filho M, Pinto e, Silva JC. Dilatation of the urinary tract during pregnancy: proposal of a curve of maximal caliceal diameter by gestational age. Am J Obstet Gynecol 1998;178: 1082–6.

[3] Rasmussen PE, Nielsen FR. Hydronephrosis during pregnancy: a literature survey. Eur J Obstet Gynecol Reprod Biol 1988;27:249–59.

[4] Marchant DJ. Effects of pregnancy and progestational agents on the urinary tract. Am J Obstet Gynecol 1972;112:487–501.

[5] Au KKL, Woo JSK, Tang LCH, Liang ST. Aetiological factors in the genesis of pregnancy hydronephrosis. Aust N Z J Obstet Gynaecol 1985;25:248–51.

[6] Conrad KP, Karumanchi SA. Renal physiology and disease in pregnancy. In: Seldin and Giebisch's the kidney: physiology and pathophysiology. Elsevier; 2012. p. 2689–761.

[7] Mahendru AA, Everett TR, Wilkinson IB, Lees CC, McEniery CM. A longitudinal study of maternal cardiovascular function from preconception to the postpartum period. J Hypertens 2014;32:849–56.

[8] Meah VL, Cockcroft JR, Backx K, Shave R, Stöhr EJ. Cardiac output and related haemodynamics during pregnancy: a series of meta-analyses. Heart 2016;102:518–26.

[9] Cheung KL, Lafayette RA. Renal physiology of pregnancy. Adv Chron Kidney Dis 2013;20:209–14.

[10] Irani RA, Xia Y. The functional role of the renin–angiotensin system in pregnancy and preeclampsia. Placenta 2008;29:763–71.
[11] Gant NFCSWPMPC. The nature of pressor responsiveness to angiotensin II in human pregnancy. Obstet Gynecol 1974;43(6):854–60.
[12] Taufield PA, Müller FB, Laragh JH. Hypertension in pregnancy. In: Hypertension in kidney disease. Dordrecht: Springer Netherlands; 1986. p. 147–69.
[13] Stennett AK, Qiao X, Falone AE, Koledova Vv, Khalil RA. Increased vascular angiotensin type 2 receptor expression and NOS-mediated mechanisms of vascular relaxation in pregnant rats. Am J Physiol Heart Circ Physiol 2009;296:H745–55.
[14] Beers K, Patel N. Kidney physiology in pregnancy. Adv Chron Kidney Dis 2020;27:449–54.
[15] Clark SL, Cotton DB, Lee W, et al. Central hemodynamic assessment of normal term pregnancy. Am J Obstet Gynecol 1989;161:1439–42.
[16] Wiles K, Stillman IE, Conrad KP. The kidney in normal pregnancy and preeclampsia. In: Chesley's hypertensive disorders in pregnancy. Elsevier; 2022. p. 289–334.
[17] Davison JM, Noble MCB. Serial changes in 24 hour creatinine clearance during normal menstrual cycles and the first trimester of pregnancy. BJOG 1981;88:10–7.
[18] Lopes van Balen VA, Gansewinkel TAG, Haas S, et al. Maternal kidney function during pregnancy: systematic review and meta-analysis. Ultrasound Obstet Gynecol 2019;54:297–307.
[19] Baylis C. The mechanism of the increase in glomerular filtration rate in the twelve-day pregnant rat. J Physiol 1980;305:405–14.
[20] Conrad KP, Jeyabalan A, Danielson LA, Kerchner LJ, Novak J. Role of relaxin in maternal renal vasodilation of pregnancy. Ann N Y Acad Sci 2005;1041:147–54.
[21] Leo CH, Jelinic M, Ng HH, et al. Vascular actions of relaxin: nitric oxide and beyond. Br J Pharmacol 2017;174:1002–14.
[22] Harel Z, McArthur E, Hladunewich M, et al. Serum creatinine levels before, during, and after pregnancy. JAMA 2019;321:205.
[23] Lindheimer MD, Barron WM, Davison JM. Osmoregulation of thirst and vasopressin release in pregnancy. Am J Physiol Ren Physiol 1989;257:F159–69.
[24] Conrad KP. Renal hemodynamics during pregnancy in chronically catheterized, conscious rats. Kidney Int 1984;26:24–9.
[25] Davison JM, Shiells EA, Philips PR, Lindheimer MD. Serial evaluation of vasopressin release and thirst in human pregnancy. Role of human chorionic gonadotrophin in the osmoregulatory changes of gestation. J Clin Invest 1988;81:798–806.
[26] Danielson LA, Sherwood OD, Conrad KP. Relaxin is a potent renal vasodilator in conscious rats. J Clin Invest 1999;103:525–33.
[27] Aggarwal S., Brown M.A. Renal physiology and complications in Normal pregnancy. In: Section VIII Pregnancy and Renal Disease Renal Physiology in Normal Pregnancy Mechanisms of Increased Glomerular Filtration Rate Renin-Angiotensin-Aldosterone System.
[28] Olausson H, Goldberg GR, Ann Laskey M, Schoenmakers I, Jarjou LMA, Prentice A. Calcium economy in human pregnancy and lactation. Nutr Res Rev 2012;25:40–67.
[29] Bardicef M, Bardicef O, Sorokin Y, et al. Extracellular and intracellular magnesium depletion in pregnancy and gestational diabetes. Am J Obstet Gynecol 1995;172:1009–13.
[30] Martin AC, Brown MA. Could uric acid have a pathogenic role in pre-eclampsia? Nat Rev Nephrol 2010;6:744–8.
[31] Lam C, Lim K-H, Kang D-H, Karumanchi SA. Uric acid and preeclampsia. Semin Nephrol 2005;25:56–60.
[32] Kattah A, Milic N, White W, Garovic V. Spot urine protein measurements in normotensive pregnancies, pregnancies with isolated proteinuria and preeclampsia. Am J Physiol Regul Integr Comp Physiol 2017;313:R418–24.
[33] Hypertension in pregnancy. Obstet Gynecol 2013;122:1122–31.

CHAPTER 2

Prenatal care, risk assessment, and counseling

Ahsina Jahan Lopa[1], Payel Bose[2] and Ahsan Ahmed[3]
[1]Department of ICU and Emergency, Brahmanbaria ICU Specialized Hospital, Dhaka, Bangladesh; [2]Critical Care Medicine, Medica Superspecialty Hospital, Kolkata, West Bengal, India; [3]Critical Care Medicine, KPC Medical College and Hospital, Kolkata, West Bengal, India

Chronic kidney disease (CKD) is reported to affect 3% of women between the ages of 20 and 39 years [1]. Until the 1980s, opinions conflicted concerning the reciprocal influence of kidney disease on the outcome of pregnancy and of pregnancy on the natural course of maternal renal disease. Several studies [2,3] and researches have considerably clarified the problem, and it was possible to identify those factors that influence fetal and maternal prognosis, thereby providing a basis for proper preconception counseling and optimal management of pregnancy [4].

The incidence of CKD in pregnancy is increasing considerably due to increasing maternal age and obesity. Women with CKD have an increased risk of antenatal complications including preeclampsia, preterm delivery, fetal growth restriction (FGR), and an accelerated loss of maternal kidney function compared to women without CKD, although a successful pregnancy is feasible for most women [5]. A systematic review reports that women with CKD have a twofold increased risk of adverse fetal outcomes and a fivefold risk of an adverse maternal event [6].

Preconceptional counseling for women with CKD allows adequate preparation for pregnancy, both physically and psychologically, as many women may be unaware that their condition has any implications for fetal or maternal health. Preconceptual review evaluates renal disease with associated comorbidities and optimize medication and planning of antenatal care.

Multidisciplinary prepregnancy counseling [7,8] should be performed for women with an estimated glomerular filtration rate (eGFR) <60 mL/min/1.73 m^2, women with CKD progression, women with uncontrolled hypertension ($>140/90$ mmHg), women with nephrotic-range proteinuria, women with active renal disease, women with lupus nephritis, women with renal transplants, and all women with previous adverse obstetric outcomes. An MDT (multidisciplinary team) is therefore recommended to facilitate informed decision-making regarding pregnancy, and to prevent and/or manage obstetric, renal, and neonatal complications that may develop. The MDT should be available prior to, during, and following pregnancy. Options for accessing the MDT include remote advice, face-to-face counseling, and the direct delivery of maternity care.

The Kidney of the Critically Ill Pregnant Woman
ISBN 978-0-443-21473-8,
https://doi.org/10.1016/B978-0-443-21473-8.00010-0

Categorization of renal disease in pregnancy

Low risk

Usually CKD 1 outside pregnancy

Serum creatinine <120 μmol/L

uPCR <100 mg/mmol

Care and delivery in local hospital, with escalation if clinical deterioration.

Significant renal function loss is unlikely so long as blood pressure and proteinuria are managed prior to conception and during pregnancy.

Moderate risk

Usually CKD 2 outside pregnancy

Serum creatinine <250 μmol/L

uPCR >100 mg/mmol under 20 weeks

Moderate increased risk of maternal or fetal mortality or morbidity.

Refer to tertiary unit to be seen before 16 weeks for management plan.

High risk

Diagnosis of CKD 3 or worse outside of pregnancy.

Lupus nephritis or vasculitis with CKD 2 or worse and or other organ involvement—refer to the LIPS clinic.

Previous flare of CKD in pregnancy.

Renal transplant/renal + pancreas transplant.

On dialysis/requiring dialysis.

New serum Creatinine >250 μmol/L.

New uPCR >300 mg/mmol.

Prepregnancy counseling and optimization for pregnancy

Prepregnancy counseling (by an MDT including a consultant obstetrician and nephrologist or expert physician) should be offered to all the women with kidney disease considering pregnancy.

Women with known or suspected inheritable renal diseases (like autosomal dominant and recessive forms of polycystic kidney disease, Alport syndromes, Fabry disease, and cystinosis) should be offered genetic counseling including inheritance risk, prognosis, and intervention options including preimplantation genetic diagnosis (PGD).

Prepregnancy counseling is required for the optimization of maternal and neonatal outcomes in women with CKD, which may include:

stabilizing disease activity in advance of pregnancy on minimized doses of pregnancy-appropriate medications;

optimizing blood pressure control (<140/90 mmHg) on pregnancy-appropriate medications;
optimizing glycemic control in women with diabetes mellitus;
minimizing risk of exposure to teratogenic medications;
making a treatment plan in the event of hyperemesis or disease exacerbation/relapse during pregnancy.

Medication

Prescribing any medication in pregnancy should involve balancing the risks to the women of uncontrolled disease, with any real or theoretical perceived harm to the fetus. Inappropriate cessation or failure to initiate therapy when clearly indicated can be more harmful than judicious use to maintain maternal health. The woman should be involved in discussions about medication in pregnancy, which should ideally take place before the pregnancy as part of prepregnancy counseling [9].

Medications considered safe in pregnancy include: calcium channel blockers (including nifedipine and amlodipine), methyldopa, hydralazine, labetalol, propranolol, prednisolone, azathioprine, tacrolimus, cyclosporine, hydroxychloroquine, aspirin and heparin.

Metformin can be used in pregnancy for women with a prepregnancy eGFR > 30 mL/minutes/1.73 m^2 and stable renal function during pregnancy.

Medications contraindicated in pregnancy include:

Angiotensin-converting enzyme (ACE) inhibitors (captopril, enalapril) and angiotensin II receptor blockers (candesartan, losartan). These medications are associated with an increased risk of congenital malformations and fetal renal dysfunction. For women with proteinuric kidney disease (including women with diabetic nephropathy) where there is a significant benefit to these treatments, a plan should be documented to discontinue these medications once a pregnancy is confirmed (ideally <9 weeks gestation).

Mycophenolate mofetil, methotrexate and cyclophosphamide are teratogenic and should be avoided in pregnancy. They should be discontinued 3—6 months before pregnancy. Ideally women should be switched to an alternative immunosuppressant medication with a period of at least 3 months stability.

Sirolimus and everolimus—limited evidence of safety in pregnancy and should be substituted with an alternative therapy before pregnancy where possible.

There is no evidence of teratogenicity with rituximab, but exposure in pregnancy can result in neonatal B cell depletion and long-term outcome is unknown.

There is no evidence of teratogenicity with eculizumab, and the benefits of use in pregnancy for organ threatening disease outweigh possible risk.

Information

Individualized information should be provided to the woman and her family about the risks of pregnancy to her long-term health and CKD. This counseling should be provided by a clinical team with expertise in kidney disease in pregnancy.

For women with CKD 3/5, the risk of transient or permanent deterioration of background renal function, risk of requiring dialysis, acute transplant rejection, and situations where termination of pregnancy is offered should be discussed. Education about kidney failure and the possibility of antenatal or postpartum dialysis initiation, including treatment options, modality choice and access, is therefore recommended prior to conception [10].

Information should be provided to the woman and her family regarding the potential complications of pregnancy and the need for additional antenatal surveillance. Counseling should be tailored to a woman's individual risk factors and should cover the risks of preeclampsia, fetal growth disorders, gestational diabetes, venous thromboembolism (VTE), preterm birth, and caesarean section.

A clear explanation of what a pregnancy will entail, the risks and possible outcomes, and the intensity of monitoring and the frequency of appointments should be provided.

There must be written communication with the renal team and GP including discussion about medication plans.

Termination of pregnancy

Rapid access to termination of pregnancy services should be facilitated, if for whatever reason a woman chooses this.

The assurance that clinicians will be nonjudgmental and supportive of a decision to terminate a pregnancy is important.

Prepregnancy care

Contraception

Although CKD impacts on mechanistic and psychological aspects of fertility, reducing the likelihood of spontaneous conceptions, unintended pregnancies occur. Although there are no recent data, a historical questionnaire study of 76 women with CKD revealed that despite 50% being sexually active, only 36% used contraception, and only 13% had discussed reproductive health issues with their nephrologist [11]. A systematic review of observational studies shows that unintended pregnancy is associated with an increased risk of obstetric complications, even in the absence of comorbidity [12], with important additional considerations in women with CKD including optimization of disease management prior to pregnancy, avoidance of teratogenic medication, and providing an awareness of an increased risk of adverse pregnancies outcomes.

Metaanalyses of observational studies show that active lupus nephritis is a significant risk factor for the development of maternal hypertension and preterm delivery [13,14]. The first year after transplantation carries the highest risk of rejection, is most likely to require management with teratogenic medication, and is associated with adverse pregnancy outcomes [15—17].

Advice on safe and effective contraception is offered to all women of reproductive age with kidney disease, who are taking teratogenic medication, have active glomerulonephritis, are within 1 year of renal transplantation or acute graft rejection, and for any woman who does not wish to conceive.

Progesterone only-pill, a progesterone subdermal implant, and the progesterone intrauterine system are safe and effective for women with CKD.

All estrogen-containing contraceptives confer risk of hypertension, VTE, arterial thrombosis, and cervical cancer [18]. These risks are particularly relevant for women with CKD with coexisting chronic hypertension and those known to be at increased risk of vascular disease, VTE (due to antiphospholipid antibodies or nephrotic syndrome), or cervical neoplasia in the context of immunosuppression.

Fertility

Women with kidney disease should be referred for prepregnancy counseling before receiving assisted reproduction.

Cohort studies show that cyclophosphamide causes age and dose-dependent gonadotoxicity in women with systemic lupus erythematosus [19,20] and a diminished ovarian reserve (quantified by longitudinal serum anti-Mullerian hormone (AMH) concentrations) in women with granulomatosis with polyangiitis [21]. Fertility preservation should therefore be considered for women of childbearing age receiving cyclophosphamide and women who have had previous treatment with cyclophosphamide should be investigated early for infertility.

Fertility preservation techniques will depend upon the urgency of treatment of the underlying condition and availability. Cryopreservation of oocytes and gametes can be undertaken, but this usually requires ovarian stimulation, which will typically delay cyclophosphamide administration and, given the immunomodulatory role of estrogen believed to underlie the female predominance of lupus, carries a theoretical risk of lupus flare. Natural cycle in vitro fertilization (IVF) negates the need for ovarian stimulation [22]. However, pregnancy rates with natural cycle IVF are lower compared to stimulated cycles and natural cycle retrieval is not recommended for women without CKD [23].

Luteinizing hormone-releasing hormone analogs (LHRHa)/gonadotrophin-releasing hormone agonists (GnRHa) can be used to inhibit the hypothalamic—pituitary—ovarian axis, leading to a protective reduction in ovarian blood flow for the duration of cyclophosphamide treatment. Data on the use of LHRH/GnRHa in women with CKD are limited.

Single-embryo transfer is performed to reduce risk of complications associated with multifetal pregnancies in women with CKD like preterm delivery, growth restriction, neonatal unit admission, weight discordance, perinatal mortality, and neonatal mortality [24].

Pregnancy care

Assessment of renal function in pregnancy

Renal function in pregnancy is assessed using serum creatinine concentrations as eGFR is not valid for use in pregnancy.

Due to increased plasma flow and dynamic changes infiltration fraction during pregnancy [25], glomerular filtration increases up to 50% with a consequent fall in serum creatinine concentrations [26]. In addition, the dynamic nature of gestational and immediate postpartum change in renal function means that steady state cannot be presumed, prohibiting the use of eGFR. Quantification of GFR by creatinine clearance in pregnancy is unreliable and impractical [27].

Antenatal care

Fetal screening

Women with CKD should be offered usual trisomy screening. If they have abnormal renal function, the pretest counseling for combined screening using blood markers should include discussion of a potentially increased false positive rate as the multiple of the median may be increased for beta human chorionic gonadotrophin, though not for pregnancy-associated plasma protein-A [28]. Other screening options such as the noninvasive prenatal testing of cell-free fetal DNA, either as a first line or subsequent to combined screening, currently depend upon local availability. Women with CKD and abnormal renal function should be referred to a specialist Fetal Medicine Unit for interpretation of positive results.

In light of the known association between steroids and calcineurin inhibitors with impaired glucose metabolism [29], screening for gestational diabetes should be arranged for women taking these medications.

Women with CKD exposed to teratogenic drugs in the first trimester are referred to a specialist fetal medicine unit.

Pregnant women with CKD should have scans to assess fetal growth and well-being in the third trimester.

Preeclampsia prophylaxis

Women with CKD (including kidney donors) should be offered low-dose aspirin (75–150 mg) in pregnancy to reduce the risk of preeclampsia.

Uterine artery Doppler ± placental biometry should be offered at 22–24 weeks. In women with normal (<95th centile) uterine artery Doppler, fetal biometry >10th centile and normal placental biometry (i.e., a 'negative placental screen'), and stable CKD 1/2, routine fetal growth surveillance may be delayed until >30 weeks (unless there is a history of a previous preterm birth (<32 weeks)).

In women with stable CKD 1/2, stable renal function and a negative placental screen, fetal growth surveillance should be carried out every 4 weeks unless signs of placental disease are suspected (abnormal growth trajectory, oligohydramnios, abnormal umbilical artery Doppler). In women with CKD 3 or worse, routine surveillance (usually four weekly) is indicated from 26 to 28 weeks.

In women with a positive placental screen at 22–24-week screen, more frequent ultrasound surveillance should be offered (usually two to three weekly).

In women with abnormal umbilical artery Dopplers and/or severe early onset FGR (<third centile) < 32 weeks, fetal surveillance by a specialist obstetrician should be offered (see FGR pathway). If this cannot be facilitated locally, referral to a specialist unit who can provide advanced surveillance (including fetal Dopplers) should be offered.

Blood pressure management

The target blood pressure during pregnancy for women with CKD is 135/85 mmHg or less.

Antihypertensive treatment in women with CKD is continued in pregnancy unless systolic blood pressure is consistently <110 mmHg systolic, or diastolic blood is pressure consistently <70 mmHg diastolic BP, or there is symptomatic hypotension.

Labetalol, nifedipine, and methyldopa can be used to treat hypertension in pregnancy.

Placental growth factor-based testing (e.g., Triage placental growth factor (PlGF) test or the Elecsys immunoassay soluble fms-like tyrosine kinase 1 (sFlt-1)/PlGF ratio) is used alongside standard clinical assessment and subsequent clinical follow-up, to help rule-out preeclampsia in women presenting with suspected preeclampsia between 20 and 34 weeks [30,31].

Venous thromboembolism

Women with nephrotic-range proteinuria (uPCR > 300 mg/mmol or ACR > 250 mg/mmol) should be offered thromboprophylaxis with low molecular weight heparin in pregnancy and the postpartum period unless there is a specific contraindication including risk of labor or active bleeding. Royal College of Obstetricians and Gynecologists (RCOG) states that all pregnant women undergo a documented assessment of risk factors for VTE [32].

Nonnephrotic range proteinuria in pregnancy is a risk factor for thrombosis and thromboprophylaxis with low molecular weight heparin should be considered in the presence of additional risk factors.

Anemia

Pregnant women with CKD should be given parenteral iron and erythropoietin stimulating agents if indicated.

No formal recommendation has been made for the use of hypoxia-inducible factor (HIF) activators in pregnancy.

However, the small size of these molecules potentially enables placental transfer, and HIF has multiple direct and indirect effects on developmental and physiological processes [33]; hence, the use of this agent has not be considered in pregnancy or lactation as of now.

Bone health

Vitamin D deficiency is estimated to affect 13%–64% of pregnant women [34] and is associated with increased incidences of preeclampsia and gestational diabetes.

Women with CKD who are vitamin D deficient should be given vitamin D supplementation in pregnancy. Calcimimetics and phosphate binders should be discontinued in pregnancy.

Specific conditions

Renal transplantation

Pregnancy rates are lower in women of reproductive age with renal transplants compared to the general population [35,36]. It is unclear whether this is due to reduced fertility or patient choice. Women with renal transplants usually have successful pregnancy outcomes, but maternal and neonatal complications remain higher compared to the general population.

Standard immunosuppression regimens routinely include mycophenolate mofetil in the first year after transplantation [37]. Due to the teratogenicity of mycophenolate mofetil, switching to alternative agents is recommended before pregnancy, thus at least a year after transplantation is advised before attempting to conceive.

Other factors to consider regarding timing of pregnancy include recent episodes of rejection, stability and level of graft function, presence of cytomegalovirus, maternal age, diabetes, and blood pressure control, but direct evidence regarding impact of these factors on pregnancy and graft outcomes is limited.

Transplant complications (infection, rejection, urological) are more during the first year posttransplantation, and more aggressive immunosuppressive treatment is used during this period. Women with renal transplants should wait until their kidney function is stable on medications safe in pregnancy before conceiving, which is usually more than 1 year after transplantation. European Best Practice Guidelines (2001) recommended a delay of 24 months between transplantation and conception [38], but American guidelines (2005) subsequently advised 12 months if stable graft function [39].

The miscarriage rate and the incidence of congenital anomalies are similar to the general population. Potential pregnancy complications in kidney transplant recipients include hypertension, preeclampsia, FGR (20%–40%), preterm delivery (45%–60%), increased risk of infections due to immunosuppression and anemia. Approximately 10%–18% of women will have a temporary or permanent deterioration of kidney function.

Acute rejection in pregnancy occurs in 9%–14% of women, but the incidence of serious episodes of rejection is 5%, which is similar to the rates observed in nonpregnant transplant patients.

Immunosuppressant therapy should be managed and monitored frequently in conjunction with the specialist transplant team.

Women with kidney–pancreas transplants, kidney–liver transplants, and dual kidney transplants should be managed during pregnancy and delivery by an MDT including transplant physicians and surgeons, at a transplant center. Dual organ transplants are associated with higher rates of adverse pregnancy outcomes [40].

Dialysis

Women established on dialysis prior to pregnancy should receive prepregnancy counseling including the options of postponing pregnancy until transplantation (when feasible) and the need for long frequent dialysis prior to and during pregnancy.

In pregnancy, concern regarding the fetotoxicity of urea is likely to precede maternal indications for dialysis, which are the same as outside of pregnancy: refractory hyperkalemia, acidosis and/or fluid overload, and uremic symptoms that impact upon daily living [41].

Metaanalysis reveals an association between increased dialysis provision and improved fertility and pregnancy outcomes [42,43].

Hemodialysis should be initiated in pregnancy when the maternal urea concentration is 17–20 mmol/L and the risks of preterm delivery outweigh those of dialysis initiation. Gestation, renal function trend, fluid balance, biochemical parameters, blood pressure, and uremic symptoms should also be considered.

Women receiving hemodialysis during pregnancy should have dialysis dose prescribed on the basis of residual renal function, aiming for a predialysis urea <12.5 mmol/L. Expert consensus is that clinical assessment of the ultrafiltration target is performed at least weekly, in order to accommodate anticipated weight gain in pregnancy of 300 g/week during the second trimester and 300–500 g/week in the third trimester [44], with a postdialysis blood pressure target <140/90 mmHg, while avoiding intradialytic hypotension <120/70 mmHg [45].

Women established on peritoneal dialysis prior to pregnancy should convert to hemodialysis during pregnancy.

The provision of long, frequent hemodialysis has implications for electrolyte and nutritional provision, and pregnant women on dialysis should have access to nutritional assessment and dietary counseling. Expert consensus is that diet of women receiving long, frequent hemodialysis should be unrestricted and rich in protein (1.5–1.8 g/kg IBW/day). Electrolytes including magnesium and calcium-phosphate balance should be monitored every 1–2 weeks. Dialyzate concentrations of potassium, calcium, and phosphate may need to be increased. Magnesium supplementation may be required. Dialyzate losses of folic acid and water-soluble vitamins need to be considered, with increased supplementation as required, including high-dose folic acid (5 mg) prepregnancy (whenever possible) and in the first trimester.

Lupus nephritis and vasculitis

The MDT should oversee the care of all women with lupus nephritis during pregnancy due to the complexities of diagnosis, combined with the need for early recognition and timely therapy to maintain maternal and fetal health. Distinguishing a flare of lupus nephritis from preeclampsia can be challenging [46].

Women with lupus nephritis have a higher risk of pregnancy complications compared to women with a comparable level of renal function due to another cause.

Women with lupus or vasculitis should be advised to wait until their disease is quiescent for at least 6 months before conceiving.

Women with active lupus, hypertension, proteinuria, reduced kidney function, or positive lupus anticoagulant should be advised that pregnancy has a greater likelihood of complication requiring preterm birth.

All women with lupus should be advised to take hydroxychloroquine in pregnancy to improve neonatal outcomes unless it is contraindicated.

Lupus disease activity should be monitored at least monthly during pregnancy.

Women who are positive for anti-Ro (SSA) or anti-La (SSB) antibodies should have the fetal heart monitored weekly after 16 weeks gestation in view of the increased 1%–2% risk of heart block. Women should be referred for a fetal echo if the heart rate is less than 110 bpm.

Women with antiphospholipid syndrome and a history of a confirmed thromboembolic event or previous adverse obstetric outcome (excluding recurrent early fetal loss) should receive low molecular weight heparin in pregnancy and for 6 weeks postpartum. Steroids, azathioprine, calcineurin inhibitors, intravenous immunoglobulin, and plasma exchange can be used to treat lupus in pregnancy.

The combination of tacrolimus and steroids is diabetogenic and women taking these drugs in isolation or combination should be screened for gestational diabetes.

Diabetic nephropathy

Diabetic nephropathy in pregnancy is associated with an increased risk of adverse outcomes including pregnancy loss, congenital malformation, preeclampsia, preterm delivery, growth restriction, and neonatal unit admission [47,48].

Prepregnancy counseling of women with diabetes is associated with improved glycemic control prior to pregnancy and a reduction in rates of spontaneous pregnancy loss and congenital malformations [49].

Proteinuria increases in pregnancy in the majority of women with diabetic nephropathy, including progression to the nephrotic range [50].

Women with diabetic nephropathy should be appropriately optimized with respect to blood glucose, blood pressure, and proteinuria, and they should continue angiotensin converting enzyme inhibitors until conception, with regular pregnancy testing during attempts to conceive.

Urinary tract infection

Asymptomatic bacteriuria is estimated to occur in 2%–7% of pregnancies with a risk of progression to acute pyelonephritis if untreated. Screening of all pregnant women for asymptomatic bacteriuria is therefore recommended [51].

Women with reflux nephropathy, congenital anomalies of the kidneys and urinary tract (CAKUT), women with CKD taking immunosuppression, and women with a history of recurrent urinary tract infection (UTI) should be offered antibiotic prophylaxis during pregnancy after a single UTI in pregnancy, including asymptomatic bacteriuria.

Penicillins, cephalosporins, fosfomycin, trimethoprim (not in first trimester), and nitrofurantoin (not at the end of pregnancy, not in glucose-6-phosphate dehydrogenase deficiency and ineffective if prepregnancy eGFR <45 mL/minutes/1.73 m^2) can be used.

Reflux nephropathy and congenital abnormalities of the kidney and urinary tract

Women with previous history of bladder surgery (reimplantation of ureter, bladder reconstruction, etc.) should be reviewed by a urologist (with expertise in bladder reconstruction) during pregnancy.

The abnormalities in the fetal kidneys and/or urinary tract detected antenatally should be discussed with fetal medicine and pediatric nephrology specialists to determine appropriate neonatal management.

Conclusion

Women with CKD are a diverse group of individuals with a spectrum of pregnancy outcomes, from those with a minimal increase in risk of preeclampsia to those very unlikely to have a normal pregnancy course. One of the most useful guides to pregnancy outcome is obstetric history, as future pregnancies often mirror previous pregnancies.

In conclusion, the present state of our knowledge supports a more optimistic view of pregnancy in patients with renal disease than in the recent past, especially with respect to fetal outcome [52]. Multidisciplinary management, preferably in a tertiary care facility, with close coordination between the nephrologist and the obstetrician from the beginning of gestation is the key factor to optimize the issue of such high-risk pregnancies. In view of the continuing progress in obstetrics and neonatology, as well as in our knowledge of factors governing fetal development in a hostile hypertensive and/or uremic milieu, one may expect that an increasing number of women in the near future will be able to obtain motherhood despite underlying renal disease.

References

[1] Coresh J, Selvin E, Stevens LA, et al. Prevalence of chronic kidney disease in the United States. JAMA 2007;298:2038–47.

[2] Ungers P, Forget D, Henry-Amar M, Albouze G, Fournier P, Vischer U, Droz D, Noel LH, Grunfeld JP. Chronic kidney disease and pregnancy. Adv Nephrol 1986;15:103–41.

[3] Packham DK, North RA, Fairley KF, Kioss M, Whitworth JA, Kincaid-Smith P. Primary glomerulonephritis and pregnancy. Q J Med 1989;71:537–53.

[4] Lindheimer MD, Katz AT. Gestation in women with kidney disease: prognosis and management. Baillieres Clin Obstet Gynaecol 1994;8:387–404.

[5] Piccoli GB, Cabiddu G, Attini R, et al. Risk of adverse pregnancy outcomes in women with CKD. J Am Soc Nephrol 2015;26:2011–22.

[6] Piccoli GB, Conijn A, Attini R, Biolcati M, Bossotti C, Consiglio V, et al. Pregnancy in chronic kidney disease: need for a common language. Evid Based Med 2011;4:17.

[7] Knight M, Tuffnell D. A view from the UK: the UK and Ireland confidential enquiry into maternal deaths and morbidity. Clin Obstet Gynecol 2018;61:347–58.

[8] Nair M, Nelson-Piercy C, Knight M. Indirect maternal deaths: UK and global perspectives. Obstet Med 2017;10:10–5.

[9] Götestam Skorpen C, Hoeltzenbein M, Tincani A, et al. The EULAR points to consider for use of antirheumatic drugs before pregnancy, and during pregnancy and lactation. Ann Rheum Dis 2016; 75:795–810.

[10] National Kidney Foundation. KDOQI clinical practice guideline for hemodialysis adequacy: 2015 update. Am J Kidney Dis 2015;66:884–930.

[11] Holley JL, Schmidt RJ, Bender FH, et al. Gynecologic and reproductive issues in women on dialysis. Am J Kidney Dis 1997;29:685–90.

[12] Shah PS, Balkhair T, Ohlsson A, et al. Intention to become pregnant and low birth weight and preterm birth: a systematic review. Matern Child Health J 2011;15:205–16.

[13] Smyth A, Oliveira GHM, Lahr BD, et al. A systematic review and metaanalysis of pregnancy outcomes in patients with systemic lupus erythematosus and lupus nephritis. Clin J Am Soc Nephrol 2010;5: 2060–8.

[14] Bundhun PK, Soogund MZS, Huang F. Impact of systemic lupus erythematosus on maternal and fetal outcomes following pregnancy: a meta-analysis of studies published between years 2001–2016. J Autoimmun 2017;79:17–27.

[15] Armenti VT, Radomski JS, Moritz MJ, et al. Report from the national transplantation pregnancy registry (NTPR): outcomes of pregnancy after transplantation. Clin Transpl 2004:103–14.

[16] Stratta P, Canavese C, Giacchino F, et al. Pregnancy in kidney transplantation: satisfactory outcomes and harsh realities. J Nephrol 2003;16:792–806.

[17] Deshpande NA, James NT, Kucirka LM, et al. Pregnancy outcomes in kidney transplant recipients: a systematic review and meta-analysis. Am J Transplant 2011;11:2388–404.

[18] Brynhildsen J. Combined hormonal contraceptives: prescribing patterns, compliance, and benefits versus risks. Ther Adv Drug Saf 2014;5:201–13.

[19] Ioannidis JPA, Katsifis GE, Tzioufas AG, et al. Predictors of sustained amenorrhea from pulsed intravenous cyclophosphamide in premenopausal women with systemic lupus erythematosus. J Rheumatol 2002;29:2129–35.
[20] Boumpas DT, Austin HA, Vaughan EM, et al. Risk for sustained amenorrhea in patients with systemic lupus erythematosus receiving intermittent pulse cyclophosphamide therapy. Ann Intern Med 1993; 119:366–9.
[21] Clowse MEB, Copland SC, Hsieh T-C, et al. Oral cyclophosphamide therapy diminishes ovarian reserve in women with granulomatosis with polyangiitis. Arthritis Care Res 2011;63:1777.
[22] Elizur SE, Chian RC, Pineau CA, et al. Fertility preservation treatment for young women with autoimmune diseases facing treatment with gonadotoxic agents. Rheumatology 2008;47:1506–9.
[23] National Institute for Health and Care Excellence. Fertility problems: assessment and treatment [CG156]. 2013. Available at: https://www.nice.org.uk/guidance/cg156. [Accessed 22 May 2019].
[24] Piccoli GB, Arduino S, Attini R, et al. Multiple pregnancies in CKD patients: an explosive mix. Clin J Am Soc Nephrol 2013;8:41–50.
[25] Odutayo A, Hladunewich M. Obstetric nephrology: renal hemodynamic and metabolic physiology in normal pregnancy. Clin J Am Soc Nephrol 2012;7:2073–80.
[26] Davison JM, Dunlop W. Renal hemodynamics and tubular function normal human pregnancy. Kidney Int 1980;18:152–61.
[27] Ahmed SB, Bentley-Lewis R, Hollenberg NK, et al. A comparison of prediction equations for estimating glomerular filtration rate in pregnancy. Hypertens Pregnancy 2009;28:243–55.
[28] Valentin M, Muller F, Beaujard MP, et al. First-trimester combined screening for trisomy 21 in women with renal disease. Prenat Diagn 2015;35:244–8.
[29] Hwang JL, Weiss RE. Steroid-induced diabetes: a clinical and molecular approach to understanding and treatment. Diabetes Metab Res Rev 2014;30:96–102.
[30] National Institute for Health and Care Excellence. PlGF-based testing to help diagnose suspected preeclampsia (Triage PlGF test, Elecsys immunoassay sFlt-1/PlGF ratio, DELFIA Xpress PlGF 1-2-3 test, and BRAHMS sFlt-1 Kryptor/BRAHMS PlGF plus Kryptor PE ratio) [DG23]. 2016. Available at: https://www.nice.org.uk/guidance/dg23. [Accessed 22 May 2019].
[31] Duhig KE, Myers J, Seed PT, et al. Placental growth factor testing to assess women with suspected preeclampsia: a multicentre, pragmatic, steppedwedge cluster-randomised controlled trial. Lancet 2019; 393:1807–18.
[32] Royal College of Obstetricians and Gynaecologists. Reducing the risk of venous thromboembolism during pregnancy and the puerperium: Greentop guideline No. 37a. 2015. Available at: https://www.rcog.org.uk/globalassets/documents/guidelines/gtg-37a.pdf. [Accessed 22 May 2019].
[33] Maxwell PH, Eckardt K-U. HIF prolyl hydroxylase inhibitors for the treatment of renal anaemia and beyond. Nat Rev Nephrol 2016;12:157–68.
[34] Royal College of Obsetricians and Gynaecologists. Vitamin D in pregnancy. Scientific Impact Paper 2014;43. Available at: https://www.rcog.org.uk/globalassets/documents/guidelines/scientific-impact-papers/vitamin_d_sip43_june14.pdf. [Accessed 22 May 2019].
[35] Gill JS, Zalunardo N, Rose C, et al. The pregnancy rate and live birth rate in kidney transplant recipients. Am J Transplant 2009;9:1541–9.
[36] Levidiotis V, Chang S, McDonald S. Pregnancy and maternal outcomesamong kidney transplant recipients. J Am Soc Nephrol 2009;20:2433–40.
[37] National Institute for Health and Care Excellence. Immunosuppressive therapy for kidney transplant in adults [TA481]. 2017. Available at: https://www.nice.org.uk/guidance/ta481. [Accessed 22 May 2019].
[38] European Best Practice Guidelines Expert Group on Renal Transplantation. European best practice guidelines for renal transplantation. Section IV: Long-term management of the transplant recipient. IV. 10. Pregnancy in renal transplant recipients. Nephrol Dial Transplant 2002;17(Suppl. 4):50–5.
[39] McKay DB, Josephson MA, Armenti VT, et al. Reproduction and transplantation: report on the AST consensus conference on reproductive issue.
[40] Transplant Pregnancy Registry International. Annual transplant pregnancy registry international report 2017. [Accessed 22 May 2019].

[41] National Institute for Health and Care Excellence. Renal replacement therapy and conservative management [NG107]. 2018. Available at: https://www.nice.org.uk/guidance/ng107. [Accessed 22 May 2019].
[42] Piccoli GB, Minelli F, Versino E, et al. Pregnancy in dialysis patients in the new millennium: a systematic review and meta-regression analysis correlating dialysis schedules and pregnancy outcomes. Nephrol Dial Transplant 2016;31:1915–34.
[43] Sachdeva M, Barta V, Thakkar J, et al. Pregnancy outcomes in women on hemodialysis: a national survey. Clin Kidney J. 2017;10:276–81.
[44] Cabiddu G, Castellino S, Gernone G, et al. Best practices on pregnancy on dialysis: the Italian study group on kidney and pregnancy. J Nephrol. 2015;28:279–88.
[45] Hladunewich M, Schatell D. Intensive dialysis and pregnancy. Haemodial Int. 2016;20:339–48.
[46] Fischer-Betz R, Specker C, Brinks R, et al. Low risk of renal flares and negative outcomes in women with lupus nephritis conceiving after switching from mycophenolate mofetil to azathioprine. Rheumatology 2013;52:1070–6.
[47] Young EC, Pires ML, Marques LP, et al. Effects of pregnancy on the onset and progression of diabetic nephropathy and of diabetic nephropathy on pregnancy outcomes. Diabetes Metab Syndr 2011;5:137–42.
[48] Ekbom P, Damm P, Feldt-Rasmussen B, et al. Pregnancy outcome in type 1 diabetic women with microalbuminuria. Diabetes Care 2001;24:1739–44.
[49] Temple R. Preconception care for women with diabetes: is it effective and who should provide it. Best Pract Res Clin Obstet Gynaecol 2011;25:3–14.
[50] Spotti D. Pregnancy in women with diabetic nephropathy. J Nephrol 2019;32:379–88.
[51] Excellence NIFHAC. Urinary tract infection (lower): clinical knowledge summaries. 2019.
[52] Epstein FH. Pregnancy and renal disease. N Engl J Med 1996;335:277–8.

CHAPTER 3

Signs and symptoms of kidney disease in the critically ill pregnant patient

Mohammed Alabdrab Alnabi[1] and Jomana Abdullah Al Mashikas[2]
[1]Department of Critical Care Medicine, Dammam Medical Complex, Dammam, Saudi Arabia; [2]Department of Obstetrics & Gynecology, Qatif Central Hospital, Qatif, Saudi Arabia

Acute kidney injury in pregnancy

Critically ill pregnant patients with acute kidney injury may have symptoms and signs related to diminished common functions of the kidney. These include peripheral edema, hypertension, oliguria, or anuria. However, many patients will not have any clear clinical symptoms or signs, and will only be diagnosed incidentally by a raised creatinine, blood urea nitrogen (BUN), hyperkalemia, or hyponatremia, in routine labs obtained during the patient's ICU stay [1−4].

Hyperkalemia is commonly seen in pregnancy-related acute kidney injury (prAKI), and it can be life threatening and may present with cardiac arrhythmias, conduction defects, and muscle weakness.

A thorough history and physical examination are essential in determining the underlying etiology of acute kidney injury in critically ill pregnant patients. It will also help guide investigations and identify and prioritize management options.

After determining the stage of pregnancy, the most crucial step in determining the etiology of acute kidney injury is to assess the patient volume status. We generally classify patients with acute kidney injury into prerenal (hypovolemic), renal (euvolemic or hypervolemic), or postrenal (obstructive).

Hypovolemia in a critically ill pregnant patient with acute kidney injury is associated with a decrease in renal and placental perfusion. Volume depletion is commonly seen following obstetric hemorrhage and severe hyperemesis gravidarum [5,6]. Clinical signs of hypovolemia include decreasing daily weight, hypotension, tachycardia, significant postural blood pressure drop of >20 mmHg, and low jugular venous pressure (JVP) [7].

Dynamic assessments of volume status in the intensive care setting are more accurate but might be more invasive.

Volume overload is often associated with clinical signs of hypertension, peripheral edema, raised JVP, oliguria or anuria, and often evidence of left ventricular failure with pulmonary edema.

Patients with sepsis and septic shock might appear to be mildly hypovolemic or euvolemic [6], while patients with thrombotic microangiopathies (TMAs), like preeclampsia,

The Kidney of the Critically Ill Pregnant Woman
ISBN 978-0-443-21473-8,
https://doi.org/10.1016/B978-0-443-21473-8.00004-5

HELLP syndrome (Hemolysis, Elevated Liver enzymes, and Low Platelets), TTP (thrombotic thrombocytopenia purpura), or aHUS (atypical hemolytic uremic syndrome), in late pregnancy or in the postpartum period [5], are often mildly hypovolemic, despite appearing edematous, due to hypoalbuminemia and increased vascular permeability [8].

In addition to determining the patient volume status, history and physical examination play a major role as well in revealing other important symptoms and signs, such as fever, hypertension, hypotension, skin rash, arthralgia, and medication history.

Preeclampsia and HELLP syndrome are the commonest causes of prAKI [9,10].

Other conditions, like acute fatty liver of pregnancy (AFLP) and TTP/aHUS are rare causes of prAKI in late pregnancy or postpartum [5].

Other causes of AKI that might not be related to pregnancy, include glomerular disease, lupus nephritis, or interstitial nephritis.

Lastly, a postrenal cause is easy to rule out, by looking at symptoms and signs of obstructive uropathy, either bladder obstruction, or bilateral ureteric obstruction.

Prerenal causes

If a critically ill pregnant patient with acute kidney injury has clinical signs of hypovolemic shock, including hypotension, tachycardia, low JVP, cold extremities, delayed capillary refill time (>5 seconds), drowsiness, fatigue, and decreased urine output, then we should rule out the possibility of volume contraction due to dehydration, or more commonly, due to hemorrhage.

A patient in early pregnancy might be severely dehydrated due to hyperemesis gravidarum, or she might be in hemorrhagic shock due to a ruptured ectopic pregnancy.

While a patient in late pregnancy or postpartum who presents with hypovolemic shock and acute kidney injury should be suspected to have an obstetric hemorrhage until proven otherwise, whether it's antepartum hemorrhage (including placental abruption or placenta previa) or postpartum hemorrhage (PPH).

Hyperemesis gravidarum

Hyperemesis gravidarum is identified by otherwise unexplained intractable vomiting and dehydration in the first to second trimesters.

Hyperemesis gravidarum occurs in up to 2% of pregnant women. Symptoms typically start at 5–6 weeks of gestation, peak around 9 weeks, and usually subside by 16–20 weeks. However, they may persist through the third trimester in 15%–20% of patients and until delivery in 5% [11,12].

Severe vomiting and poor oral intake may result in a state of hypovolemia, postural dizziness, postural hypotension, and potentially severe hypokalemia and metabolic alkalosis. Such volume and electrolyte deficits can result in acute kidney injury and acute tubular necrosis.

Ruptured ectopic pregnancy

The most common clinical presentation of ruptured ectopic pregnancy is first-trimester vaginal bleeding (76%) and/or abdominal pain (66%) [13], although these symptoms may occur in many other conditions in early pregnancy. It may also present with syncope, and may result in life-threatening intraabdominal hemorrhage, hemorrhagic shock, and acute kidney injury (18%).

Clinical manifestations of ectopic pregnancy usually start 6–8 weeks after the last normal menstrual period but may also start later.

Antepartum hemorrhage

Patients with acute placental abruption classically present in late pregnancy with sudden onset of abdominal pain, back pain, and uterine contractions. It may or may not be associated with vaginal bleeding. The uterus is usually firm and may be tender with fetal parts not palpable.

However, not all signs may be present simultaneously. In particular, the combination of abdominal pain, hypotension, and fetal heart rate abnormalities suggests clinically significant separation that could result in severe maternal morbidity and fetal death [14,15]. By comparison, the amount of vaginal bleeding does not correlate with the severity of placental separation. Severe abruption can have mild vaginal bleeding, because a large volume of blood can be concealed behind the placenta.

90% of women with placenta previa present with painless vaginal bleeding in the second half of pregnancy [16]. 10% of women present with uterine contractions, pain, and bleeding, similar to the presentation of placental abruption [17].

Antepartum hemorrhage can be quite severe, resulting in hemorrhagic shock, and acute kidney injury.

Postpartum hemorrhage

In a postpartum woman with symptoms and signs of hypovolemia or hypovolemic shock, one should always rule out PPH, before attributing it to other less concerning diagnoses (like vasovagal episodes or high spinal anesthesia).

A high index of suspicion is paramount, because visual quantification of blood loss might not be accurate. Maternal early warning signs criteria include tachycardia, tachypnea, hypotension, and reduced urine output. Other signs include maternal agitation, confusion, or unresponsiveness. An immediate response to these thresholds can reduce maternal morbidity [18].

Note that a large reduction in blood pressure is a late sign of severe PPH as it is generally not manifested until substantial bleeding has occurred, and up to 25% of a patient's blood volume ($\geq$1500 mL in pregnancy) can be lost before blood pressure falls and heart rate rises [19].

Delayed recognition and intervention of PPH may lead to hemorrhagic shock and the development of the potentially lethal triad of hypothermia, acidosis, and coagulopathy, leading to multi organ dysfunction, including acute kidney injury. 90% of PPH deaths occur within 4 hours of delivery [20].

Renal causes

If a critically ill patient in early pregnancy presents with clinical signs of shock, acute kidney injury, and normal volume status, suspect sepsis, most commonly septic miscarriage.

If a critically ill patient in late pregnancy or postpartum has acute kidney injury with normal or high volume status, multiple differential diagnoses should be considered. These include sepsis, TMAs, and AFLP.

Infection/sepsis

Maternal sepsis may be complicated by septic shock, which has been reported to occur in 0.04% of all deliveries [21], and acute kidney injury. This is a significant cause of maternal morbidity and mortality [19]. The reported maternal mortality rate ranges from 12% to 28% in patients with septic shock and multiple organ failure [21,22].

The most common cause of severe infection in early pregnancy is septic miscarriage. While those that occur in late pregnancy but before delivery include chorioamnionitis and pyelonephritis [22,23].

In the postpartum period, the most common severe infection is endometritis.

Early identification of sepsis can be facilitated by using various sepsis scores. The performance of these scores appears to have similar limitations as nonpregnant patients. Table 3.1 below shows the various score criteria and their sensitivity and specificity [22,24].

Table 3.1 The components & performance of various sepsis scores for early identification of sepsis in pregnancy.

Score	Criteria	Sensitivity	Specificity
Systemic inflammatory response system (SIRS)	Heart rate, respiratory rate, temperature, and leukocyte count	**0.93**	**0.63**
Quick sequential (sepsis-related) organ failure assessment (qSOFA)	Blood pressure, respiratory rate, and mental status	**0.5**	**0.95**
Modified early warning score (MEWS)	Heart rate, blood pressure, respiratory rate, temperature, mental status, and urine output	**0.82**	**0.87**
Sepsis in obstetrics score (SOS)	Heart rate, blood pressure, respiratory rate, temperature, oxygenation, leukocyte count, and lactate.	**0.89**	**0.99**

Septic miscarriage

Acute kidney injury due to septic miscarriage has become a rare complication of pregnancy due to better prenatal care [25]. Septic miscarriage can occur while products of conception are still in utero, or after pregnancy termination.

Patients with septic miscarriages commonly present with pelvic pain, abdominal pain, uterine tenderness, purulent vaginal discharge, vaginal bleeding, and fever.

Signs of septic miscarriage include high-grade fever (>38.0°C) or hypothermia, tachypnea, tachycardia, and leukocytosis or leukopenia. Pain out of proportion of physical tenderness may indicate necrotizing infection.

This may progress to severe infection, sepsis, and multiorgan dysfunction including acute kidney injury. In the setting of severe hypotension and shock, acute tubular necrosis can develop, which can take days to weeks to recover; or in severe cases result in end-stage renal disease.

Acute pyelonephritis

Acute pyelonephritis usually occurs after 20 weeks of pregnancy, but may also occur at any time during pregnancy [26].

It is usually characterized by dysuria, flank pain, nausea, and fever, and it may progress to septic shock and acute kidney injury, characterized by hypotension and tachycardia [25,27], although it might be rare to progress to acute oliguric renal failure [27].

Chorioamnionitis

Chorioamnionitis is the infection of the placenta and amniotic membranes. It occurs in 1%–2% of term and 5%–10% of preterm deliveries (ACOG).

The key diagnostic criteria of the American College of Obstetrics and Gynecology (ACOG) are as follows:

Maternal fever >39C, or 38–39C on two occasions (incidence 100%), and at least one of the following:

- Elevated maternal white blood cells.
- Fetal tachycardia >160/min (40%–70%).
- Purulent amniotic fluid.

Others clinical findings include maternal tachycardia >100/min (50%–80%) and uterine tenderness (4%–25%) [28].

Endometritis

The key clinical findings of endometritis are postpartum fever and uterine tenderness. Additional findings include excessive uterine bleeding, tachycardia, malodorous purulent lochia, headache, chills, malaise, and/or anorexia.

Severe endometritis can present with signs of sepsis, including high grade fever (although it should still be considered even without fever), tachycardia, tachypnea, and hypotension. This can progress to septic shock with multiorgan dysfunction, including acute kidney injury [29].

Thrombotic microangiopathies

TMA is a pathological definition; it relates to the presence of anemia, fragmented red blood cells (schistocytes), thrombocytopenia, raised lactate dehydrogenase (LDH), and evidence of end organ damage such as acute kidney injury, neurological, or liver involvement [5].

In pregnancy, preeclampsia is the commonest cause of TMA [30]. TTP and atypical hemolytic-uremic syndrome (aHUS) are less common causes of TMA in pregnancy and the postpartum period.

They usually present with mild to moderate AKI, but they may also present with severe AKI, with significant oliguria, anuria, or even dialysis dependence.

A combination of clinical and laboratory features is used to guide and determine the underlying etiology of TMA. In such complex cases, it is always critical to involve multidisciplinary teams early in the management of these women.

Preeclampsia

Preeclampsia is a disorder of pregnancy, presenting with new onset hypertension, that occurs in late pregnancy or postpartum, often associated with proteinuria; but sometimes present with other nonspecific signs and symptoms in the absence of proteinuria. These include persistent or severe headache, visual abnormalities, epigastric pain, altered mental status, or dyspnea.

Preeclampsia leads to AKI in 1.5%–2% of cases [31]. However, because preeclampsia itself is common, this means that it is the most prevalent cause of AKI in pregnancy as well as the commonest glomerular disease in the world [32]. AKI in preeclampsia is a marker of severity [33]. Preeclampsia with severe features may be accompanied by acute tubular necrosis [25].

The oliguria of preeclampsia does not automatically infer volume depletion. In fact, intravenous hydration in women with preeclampsia is associated with pulmonary edema and increased maternal mortality. There is no evidence that treating to a target urine output will prevent the rare development of overt renal failure and a urine output >40 mL in 4 hours is sufficient in the immediate postpartum period [34]. Invasive monitoring can be used to guide complex fluid management.

Hemolysis, elevated liver enzymes, and low platelets syndrome

HELLP syndrome is a variant of preeclampsia with severe features, which is associated with a higher incidence of AKI (3%–15%) [35]. The risk of AKI in HELLP syndrome increases if abruption, disseminated intravascular coagulopathy (DIC), sepsis, hemorrhage, or intrauterine death occur. AKI worsens the prognosis of HELLP syndrome.

AKI in HELLP usually recovers in most patients, but it can be severe enough to require renal replacement therapy [36]. HELLP may deteriorate for 48 hours postpartum, but 90% of cases will improve within <7 days [37]. Patients who fail to improve, or if thrombocytopenia and AKI predominate the clinical picture, TTP or aHUS should be considered.

Upper abdominal pain and tenderness are the most common symptom and sign [38]. Many women will have nausea, vomiting, and generalized malaise. Some women may also have headache, visual disturbance, ascites, and jaundice.

Hypertension and proteinuria are present in 85% of cases, while 15% will not have either or both of them [39].

Thrombotic thrombocytopenic purpura/atypical hemolytic uremic syndrome

Pregnancy-associated TMA occurs in 1:25,000 pregnancies [40]. Microvascular thrombosis leads to consumptive thrombocytopenia, mechanical hemolysis, and end-organ dysfunction. TMA is divided into two clinical phenotypes: TTP with primarily neurological symptoms and HUS with predominant renal dysfunction. However, these phenotypes have significant clinical overlap. The main differentiating factor is either an abnormality of ADAMTS13 (a disintegrin and metalloproteinase with a thrombospondin type I motif member 13) in TTP or complement pathway dysregulation in aHUS. TTP is predominantly a disease of late pregnancy, while aHUS is a disease of the postpartum period [41].

AKI incidence in pregnancy-associated TTP is much higher than in TTP without pregnancy, occurring in 30%–80% of cases. The incidence of AKI is even higher in aHUS, occurring in 90%–100% of cases, and 76% will progress to end-stage renal disease [42].

Both TTP and aHUS can be difficult to distinguish from HELLP. 20% of women with pregnancy-associated TTP are diagnosed with coexisting preeclampsia/HELLP [42], as thrombocytopenia, hemolysis, renal dysfunction, and neurological symptoms are part of the clinical picture of all these conditions.

Acute fatty liver of pregnancy

AFLP is an obstetric emergency occurring in 1:20,000 pregnancies [43], typically presenting in late pregnancy or early postpartum [44].

The initial symptoms are usually nonspecific, including anorexia, nausea, vomiting, abdominal pain, general malaise, and headache. Many patients may also have hypertension or proteinuria, and some may have jaundice and ascites.

This can rapidly progress into acute hepatic failure, with hypoglycemia, encephalopathy, and DIC. Most patients develop acute kidney injury (50%–60%), with 3.5% requiring renal replacement therapy, who often progress to multiorgan dysfunction [43,45]. However, most women will recover liver function and AKI within 1 week after delivery [46].

AFLP can easily be confused with HELLP, because they both share a similar clinical picture of low platelets and deranged liver function. However, women with AFLP are more likely to have hypoglycemia and hyperammonemia. Although preeclampsia/HELLP can coexist with AFLP in 20%–40% of patients [47].

Postrenal causes

Obstructive uropathy is rare in pregnancy. However, it is more likely to lead to AKI in women with kidney transplant or a single functioning kidney.

It usually presents with oliguria or anuria, flank or suprapubic pain, and moderate to severe dilatation of the collecting system on ultrasound.

If they present late, they might have signs of acute renal failure, including metabolic acidosis, volume overload, signs of hyperkalemia, or uremic encephalopathy.

Diagnosis of obstruction is difficult in pregnancy because of physiological hydronephrosis that starts usually from the second trimester without clinical significance, which may be hard to distinguish from pathological hydronephrosis.

Before looking for other diagnoses, one should rule out the most obvious cause of obstructive uropathy, which is a blocked Foley catheter. In early pregnancy, a retroflexed uterus can also lead to acute urinary retention, while in postpartum, urinary retention occurs in up to 15% of patients and has physiological, neurological, and mechanical causes [48]. Neurogenic bladder needs to be considered in women with diabetes or multiple sclerosis. Kidney stones are rare in pregnancy. Women may present with symptoms of renal colic or superimposed urosepsis (abscess or pyonephrosis).

Nonpregnancy-related causes

Sometimes, AKI in pregnancy may be caused by nonpregnancy-related conditions that may have a temporal relationship with pregnancy, or get exacerbated by the physiological state of pregnancy.

Drugs should always be considered in critically ill pregnant patients with AKI, since they are the most common cause of acute interstitial nephritis and kidney injury [49]. There are no specific signs or symptoms of drug-induced nephrotoxicity. Drugs that may be newly prescribed in pregnancy include penicillins, cephalosporins, gentamicin, vancomycin, proton-pump inhibitors, or H_2 receptor antagonists. In the postpartum period, the most widely prescribed potentially nephrotoxic drugs are the nonsteroidal antiinflammatory drugs (NSAIDs).

Hypertension, hematuria with red cell casts, and a rapidly rising serum creatinine concentration are usually due to acute glomerulonephritis or renal vasculitis. Other clues to the diagnosis of glomerulonephritis include proteinuria, volume overload (edema of the face, hands, feet, and abdomen), oliguria, nausea, and vomiting.

Arthralgias, synovitis, malar rash, mouth ulcers, and alopecia may be clues for the diagnosis of systemic lupus erythematosus. SLE is an autoimmune disease that affects women of childbearing age. 20%–50% of women with SLE will have lupus nephritis, which can present acutely for the first time during pregnancy, and in 10% of cases, can progress to acute renal failure. Lupus activity is known to increase in pregnancy [50], with 16% incidence of renal disease activation in pregnancy [51]. SLE classically presents with AKI, proteinuria, and hypertension, which makes it difficult to distinguish from preeclampsia due to the similar clinical features [52].

Pregnancy with chronic kidney disease

Women with chronic kidney disease (CKD) have reduced fertility. But if they do get pregnant, they are at an increased risk of adverse maternal and fetal events, like acute on top of chronic kidney disease, nephrotic syndrome, or hypertension, with an increase in overall mortality.

A third of the women with grade-3 CKD further deteriorated their kidney function with pregnancy, and persisted for 6 months postpartum. Women with grade-4 CKD may progress to end-stage renal disease with pregnancy. The risk may be higher with diabetic nephropathy or lupus nephritis.

Symptoms and signs of CKD include weakness, easy fatiguability, anorexia, vomiting, and mental status changes. Most pregnant women with CKD will have hypertension (25%) and increased proteinuria (50%). Anuria is rarely observed with pure CKD and indicates at least some component of AKI [53].

Women on dialysis will have significantly diminished fertility. Only about 1.5% of women on long-term dialysis become pregnant. However, pregnancy still occurs, specially during the first year of dialysis and with intensive dialysis therapy. Postkidney transplant, fertility improves, and the impact of pregnancy on graft survival is low (5.8% at 1 year and 6.9% at 5 years) [54].

References

[1] KDIGO. KDIGO clinical practice guideline for acute kidney injury. Kidney Int 2012;2:1–138.
[2] Krane NK, Hamrahian M. Pregnancy- kidney diseases and hypertension. Am J Kidney Dis February 2007;49(2):336–45.
[3] Lindheimer MD, Taler SJ, Cunningham FG. Hypertension in pregnancy. J Am Soc Hypertens 2010; 4(2):68–78.
[4] Fitzpatrick A, Mohammadi F, Jesudason S. Managing pregnancy in chronic kidney disease- improving outcomes for mother and baby Int J Women Health 2016:8–273-85.
[5] Fakhouri F, Scully M, Provot F, et al. Management of thrombotic microangiopathy in pregnancy and postpartum: report from an international working group. Blood 2020;136(19):2103–17.
[6] Szczepanski J, Griffin A, Novotny S, et al. Acute kidney injury in pregnancies complicated with preeclampsia or HELLP syndrome. Front Med 2020;7:22.
[7] McGee S, Abernethy 3rd WB, Simel DL. The rational clinical examination. Is this patient hypovolemic? JAMA 1999;281(11):1022–9.

[8] Brown MA, Zammit VC, Lowe SA. Capillary permeability and extracellular fluid volumes in pregnancy-induced hypertension. Clin Sci 1989;77(6):599–604.

[9] Mehrabadi A, Liu S, Bartholomew S, et al. Hypertensive disorders of pregnancy and the recent increase in obstetric acute renal failure in Canada: population based retrospective cohort study. BMJ 2014;349:g4731.

[10] Huang C, Chen S. Acute kidney injury during pregnancy and puerperium: a retrospective study in a single center. BMC Nephrol 2017;18(1):146.

[11] Association of Professors of Gynecology and Obstetrics. Nausea and vomiting of pregnancy. https://www.apgo.org/. (Accessed on December 05, 2016).

[12] Risk factors for hyperemesis gravidarum requiring hospital admission during pregnancy, Fell DB, Dodds L, Joseph KS, Allen VM, Butler B. Obstet Gynecol 2006;107(2 Pt 1):277.

[13] Alkatout I, Honemeyer U, Strauss A, Tinelli A, Malvasi A, Jonat W, Mettler L, Schollmeyer T. Clinical diagnosis and treatment of ectopic pregnancy. Obstet Gynecol Surv August 2013;68(8):571–81.

[14] Mei Y, Lin Y. Clinical significance of primary symptoms in women with placental abruption. J Matern Fetal Neonatal Med September 2018;31(18):2446–9. Erratum in: J Matern Fetal Neonatal Med. 2018 Nov;31(21):2932.

[15] Kasai M, Aoki S, Ogawa M, Kurasawa K, Takahashi T, Hirahara F. Prediction of perinatal outcomes based on primary symptoms in women with placental abruption. J Obstet Gynaecol Res June 2015; 41(6):850–6.

[16] Fan D, Wu S, Liu L, Xia Q, Wang W, Guo X, Liu Z. Prevalence of antepartum hemorrhage in women with placenta previa: a systematic review and meta-analysis. Sci Rep January 9, 2017;7:40320.

[17] Silver R, Depp R, Sabbagha RE, Dooley SL, Socol ML, Tamura RK. Placenta previa: aggressive expectant management. Am J Obstet Gynecol September 1, 1984;150(1):15–22.

[18] Shields LE, Wiesner S, Klein C, Pelletreau B, Hedriana HL. Use of maternal early warning trigger tool reduces maternal morbidity. Am J Obstet Gynecol April 2016;214(4):527.e1–6.

[19] Bonnar J. Massive obstetric haemorrhage. Baillieres Best Pract Res Clin Obstet Gynaecol February 2000;14(1):1–18.

[20] Prata N, Gerdts C. Measurement of postpartum blood loss. BMJ February 1, 2010;340:c555.

[21] Hensley MK, Bauer ME, Admon LK, Prescott HC. Incidence of maternal sepsis and sepsis-related maternal deaths in the United States. JAMA September 3, 2019;322(9):890–2.

[22] Bauer ME, Housey M, Bauer ST, Behrmann S, Chau A, Clancy C, Clark EAS, Einav S, Langen E, Leffert L, Lin S, Madapu M, Maile MD, McQuaid-Hanson E, Priessnitz K, Sela HY, Shah A, Sobolewski P, Toledo P, Tsen LC, Bateman BT. Risk factors, etiologies, and screening tools for sepsis in pregnant women: a multicenter case-control study. Anesth Analg December 2019;129(6):1613–20.

[23] Sheffield JS, Cunningham FG. Community-acquired pneumonia in pregnancy. Obstet Gynecol October 2009;114(4):915–22.

[24] Albright CM, Has P, Rouse DJ, Hughes BL. Internal validation of the sepsis in obstetrics score to identify risk of morbidity from sepsis in pregnancy. Obstet Gynecol October 2017;130(4):747–55.

[25] Grünfeld JP, Pertuiset N. Acute renal failure in pregnancy. Am J Kidney Dis 1987;9(4):359. 1987.

[26] Acute pyelonephritis in pregnancy: a retrospective study. Sharma P, Thapa L. Aust N Z J Obstet Gynaecol August 2007;47(4):313–5.

[27] Suppurative bacterial pyelonephritis as a cause of acute renal failure, Thompson C, Verani R, Evanoff G, Weinman E. Am J Kidney Dis 1986;8(4):271.

[28] Tita AT, Andrews WW. Diagnosis and management of clinical chorioamnionitis. Clin Perinatol June 2010;37(2):339–54.

[29] Bauer ME, Lorenz RP, Bauer ST, Rao K, Anderson FWJ. Maternal deaths due to sepsis in the state of Michigan, 1999-2006. Obstet Gynecol October 2015;126(4):747–52.

[30] Bayer G, von Tokarski F, Thoreau B, et al. Etiology and outcomes of thrombotic microangiopathies. Clin J Am Soc Nephrol 2019;14(4):557–66.

[31] Gammill HS, Jeyabalan A. Acute renal failure in pregnancy. Crit Care Med 2005;33(Suppl. 10): S372–84.

[32] Stillman IE, Karumanchi SA. The glomerular injury of preeclampsia. J Am Soc Nephrol 2007;18: 2281–4.

[33] Menzies J, Magee LA, Macnab YC, Ansermino JM, Li J, Douglas MJ, et al. Current CHS and NHBPEP criteria for severe preeclampsia do not uniformly predict adverse maternal or perinatal outcomes. Hypertens Pregnan 2007;26:447–62.
[34] von Dadelszen P, Menzies J, Gilgoff S, Xie F, Douglas MJ, Sawchuck D, et al. Evidence-based management for preeclampsia. Front Biosci 2007;12:2876–89.
[35] Gul A, Aslan H, Cebeci A, Polat I, Ulusoy S, Ceylan Y. Maternal and fetal outcomes in HELLP syndrome complicated with acute renal failure. Ren Fail 2004;26:557–62.
[36] August P. Preeclampsia: a "nephrocentric" view. Adv Chron Kidney Dis 2013;20:280–6.
[37] Neiger R, Contag SA, Coustan DR. The resolution of preeclampsia-related thrombocytopenia. Obstet Gynecol 1991;77:692–5.
[38] Sibai BM, Ramadan MK, Usta I, Salama M, Mercer BM, Friedman SA. Maternal morbidity and mortality in 442 pregnancies with hemolysis, elevated liver enzymes, and low platelets (HELLP syndrome). Am J Obstet Gynecol October 1993;169(4):1000–6.
[39] Sibai BM. Diagnosis, controversies, and management of the syndrome of hemolysis, elevated liver enzymes, and low platelet count. Obstet Gynecol May 2004;103(5 Pt 1):981–91.
[40] Matthews JH, Benjamin S, Gill DS, Smith NA. Pregnancy-associated thrombocytopenia: definition, incidence and natural history. Acta Haematol 1990;84:24–9.
[41] Fakhouri F, Roumenina L, Provot F, Sallee M, Caillard S, Couzi L, et al. Pregnancy-associated hemolytic uremic syndrome revisited in the era of complement gene mutations. J Am Soc Nephrol 2010;21:859–67.
[42] Ganesan C, Maynard SE. Acute kidney injury in pregnancy: the thrombotic microangiopathies. J Nephrol 2011;24:554–63.
[43] Knight M, Nelson-Piercy C, Kurinczuk JJ, Spark P, Brocklehurst P. A prospective national study of acute fatty liver of pregnancy in the UK. Gut 2008;57:951–6.
[44] Westbrook RH, Dusheiko G, Williamson C. Pregnancy and liver disease. J Hepatol April 2016;64(4): 933–45.
[45] Nelson DB, Yost NP, Cunningham FG. Acute fatty liver of pregnancy: clinical outcomes and expected duration of recovery. Am J Obstet Gynecol November 2013;209(5):456.e1–7.
[46] Machado S, Figueiredo N, Borges A, Sao Jose Pais M, Freitas L, Moura P, et al. Acute kidney injury in pregnancy: a clinical challenge. J Nephrol 2012;25:19–30.
[47] Liu J, Ghaziani TT, Wolf JL. Acute fatty liver disease of pregnancy: updates in pathogenesis, diagnosis, and management. Am J Gastroenterol June 2017;112(6):838–46.
[48] Chaurasia A, Tyagi K. Persistent postpartum urinary retention following vaginal delivery: a rare complication in obstetrics practice. Int J Reprod Contracept Obstet Gynecol 2013;2:475–7.
[49] Perazella MA, Markowitz GS. Drug-induced acute interstitial nephritis. Nat Rev Nephrol 2010;6:461–70.
[50] Ruiz-Irastorza G, Lima F, Alves J, Khamashta MA, Simpson J, Hughes GR, et al. Increased rate of lupus flare during pregnancy and the puerperium: a prospective study of 78 pregnancies. Br J Rheumatol 1996;35:133–8.
[51] Smyth A, Oliveira GH, Lahr BD, Bailey KR, Norby SM, Garovic VD. A systematic review and meta-analysis of pregnancy outcomes in patients with systemic lupus erythematosus and lupus nephritis. Clin J Am Soc Nephrol 2010;5:2060–8.
[52] Zhao C, Zhao J, Huang Y, et al. New-onset systemic lupus erythematosus during pregnancy. Clin Rheumatol 2013;32(6):815–22.
[53] Zhang JJ, Ma XX, Hao L, et al. A systematic review and meta-analysis of outcomes of pregnancy in CKD and CKD outcomes in pregnancy. Clin J Am Soc Nephrol 2015;10(11):1964–78.
[54] Deshpande NA, et al. Pregnancy outcomes in kidney transplant recipients: a systematic review and meta-analysis. Am J Transplant 2011;11(11):2388–404.

CHAPTER 4

Nutrition in pregnant women with renal involvement

Maria Cristina Jiménez Bazzano[1] and Elvira Gimenez Rolón[2]
[1]Department of Internal Medicine and Clinical Nutrition, Faculty of Medical Sciences, National University of Asunción, Clinicas Hospital, Asunción, Paraguay; [2]Department of Internal Medicine-Nephrology, Faculty of Medical Sciences, National University of Asunción, Clinicas Hospital, Asunción, Paraguay

Renal physiologic changes in pregnant women with kidney disease requires prompt treatment to prevent kidney damage from progressing and creating more complications.

Pregnancy in women with chronic kidney disease (CKD) has increased due to the use of erythropoietin, a greater number of women undergoing dialysis, improved forms of treatment, and nutritional status [1].

Kidney disease that develops during pregnancy is usually due to:

1. New onset glomerulonephritis
2. Lupus nephritis
3. Interstitial nephritis
4. Acute renal failure [2].

Nutritional treatment and adequate weight gain are essential for a successful pregnancy in patients with chronic renal failure (CRF) [3].

The objectives are:

Maintain a Nutritional State that allows the adequate development of the pregnancy.

Provide, through the diet, the nutrients necessary for the normal growth and development of the fetus.

Maintain homeostasis as close as possible to normal, to achieve a balance between the absorption of nutrients required by pregnancy and the catabolism of the dialysis procedure.

Carry out a monthly nutritional control and follow-up, in order to minimize the complications of pregnant women with CKD.

Promote proper metabolic control, vitamin and mineral supplements individually.

Nutritional assessment

Consider: cause and time of evolution of the kidney disease, presence of other medical conditions, data such as eating habits, appetite, intestinal transit.

Habit: tobacco, alcohol, illicit drugs.

3-day diet log.

24-hour reminder.

The Kidney of the Critically Ill Pregnant Woman
ISBN 978-0-443-21473-8,
https://doi.org/10.1016/B978-0-443-21473-8.00008-2

Na, K, Ca intake.

Nutritional scales: subjective global assessment (VGS) and Malnutrition Inflammation Score (MIS) [4].

Pregestational dry weight.

Body mass index

Skin folds, handgrip.

History of recent weight change.

Body composition analysis: impedance, DEXA bone density test.

Indirect calorimetry, rest energy expenditure equations.

Laboratory: request albumin, prealbumin, nPCR urea, creatinine clearance, 24-hour proteinuria, see Table 4.1.

Weight gain

Weight gain during pregnancy is an important indicator of energy balance and must be monitored.

It is determined by prepregnancy weight, body mass index.

The recommendations for weight gain from the Institute of Gestational Medicine range from 5 kg (in obesity) to 18 kg (in low weight).

A gestational weight gain of 13.6 kg is suggested in women of normal weight [6,7]. Recommended Weight gain is shown in Table 4.2.

Energy and macronutrient recommendations in pregnant women with CKD.

Table 4.1 Laboratory objectives to be achieved with nutritional intervention [9].

Parameter	CKD pre-dialysis	Dialysis	Transplant
Albumin	≥3.5 g/dL	≥3.5 g/dL	
BUN	≤50 mg/dL	≤40 mg/dL	≤25 mg/dL
Hemoglobin	10–14 mg/dL	10–11 mg/dL	10–14 mg/dL
Potassium	3.5–5.5 mEq/L	3.5–5.5 mEq/L	3.5–5.5 mEq/L
Phosphorus	2.5–6 mg/dL	4.5–6 mg/dL	2.5–6 mg/dL
Calcium	8.5–10.5 mg/dL	10–11 mg/dL	8.5–10.5 mg/dL
Cholesterol	<300 mg/dL	<300 mg/dL	<300 mg/dL
Glycemia pre-Prandial	60–110 mg/dL	60–126 mg/dL	60–110 mg/dL

CKD patients not undergoing dialysis have the same indication as pregnant women without CKD with nutritional counseling to limit nonnutritive sources of phosphate if necessary [5].

Table 4.2 Recommended weight gain based on body mass index (BMI).

Weight classification based on BMI	Total recommended weight gain
BMI < 19.8	12.5–18
BMI 19.8–26	11.5–16
Excess weight BMI 26–29	7–11.5
Obesity BMI 30	7

Adapted for Human energy requirements FAO/OMS/NU Roma 2004.

Energy recommendations

Chronic renal disease: resting energy expenditure (REE) is higher in adults with CKD due to renal replacement therapy, poorly controlled diabetes mellitus (DM), severe hyperparathyroidism (PTH), and inflammation.

In nondialyzed patients, REE tends to be equal to or even lower than that of healthy individuals [8,9].

Macronutrient intake [10].

Carbohydrates: 50%–60% of calories, 275 g (1100 kcal); preferably complex, slowly absorbed, to decrease triglyceride synthesis and improve glucose tolerance [11].

Fats: 30%–40%, 100 g (700 kcal), are provided in the form of mono and polyunsaturated fatty acids.

Protein: A diet of 0.6–0.9 g/kg ideal body weight/day in patient with DM, without DM metabolically stable, recommend prescribing a dietary protein intake of 1.0–1.2 g/kg [9,12,13,15,16].

Nutrition in predialysis

The initial medical nutritional therapy intervention for renal failure is described in Table 4.3 and depends on the level of renal function and maternal needs.

Protein restriction is balanced to prevent a uremic state (blood urea nitrogen [BUN] < 60 mg/dL) and prevent hyperfiltration, while at the same time preventing a decrease in albumin below 3.5 g/dL [9,14–16].

Nutrition in dialysis

Table 4.4 describes the initial medical nutrition therapy intervention needed for the pregnant woman [9,14,15,17].

Table 4.3 Nutritional recommendations for pregnant women with chronic renal failure not on dialysis [9,15,16].

Nutrient	Recommendation
Kilocalories	Basal energy expenditure × physical activity + 100–300 kcal/day. 2nd/3rd trimester or 25–45 kcal/prepregnancy weight + 100–300 kcal/day
Proteins	0.6–0.8 g/kg pregestational need + 10–16 g/day with DM. Without DM metabolically stable, recommend prescribing a dietary protein intake of 1.0–1.2 g/kg
Na–K	Individualized 2–4 g/day
Phosphorus	Individualized: 800–1200 mg/day
Calcium	Individualized: 1000–1200 mg/day

Table 4.4 Nutritional recommendations for pregnant women with chronic renal failure in dialysis [9,15,16].

Nutrient	Recommendation
Kilocalories	Individualized: Basal energy expenditure × physical activity + 100–300 kcal/day. 2nd/3rd trimester or 25–45 kcal/pregestational weight + 100–300 kcal/day Peritoneal dialysis consider kcal absorbed from the patient in dialysis
Proteins	Hemodialysis 1.2 g/kg pregestational + 10 g/day or + according to stress or metabolic need, in DM at not DM Peritoneal dialysis: 1.5 g/kg pregestational weight + 10 g/day according to stress or metabolic in DM at nor DM Objective: To maintain serum urea levels <90 mg/dL, albumin >3.5 g/dL, creatinine <5 mg/dL
Na–K	Individualized 2–4 g/day
Phosphorus	Individualized: 1200 mg/day
Calcium	Individualized: 1200 mg/day
Carnitine	330 mg/day

The main objective of the nutritional intervention is to maintain adequate protein reserve with a protein intake equal to that which a dialysis patient requires more additional pregnancy needs [18].

Dry weight assessment is performed weekly for ongoing changes in tissue and fetal weight that are occurring.

Weekly dry weight gain: 300 g.

Increase in dry weight from second trimester: 500 g/week.

Supplementation of water-soluble vitamins should be indicated. Add active vit D.

Rigorous control of interdialytic weight.

Weekly dry weight gain: 300 g.

Increase in dry weight from second trimester: 500 g/week.

Vitamins

The necessary vitamins during pregnancy can be seen in Table 4.5.

Carnitine is an important nutrient in the prenatal period because metabolism is altered in renal failure and is an important nutrient in cell metabolism [9,19].

Carnitine at a dose of 330 mg/day decreases complications of pregnancy, premature birth [20].

The amount of iron to be administered will depend on the iron stores added to iron losses on dialysis, the action of binders, and the dose of erythropoietin (EPO) administered.

Table 4.5 RDI nutrients pregnant women recommendations.

Vitamins	Dose	Potential effects on pregnancy
Vitamin A μg	1300	Excess vit A: Spontaneous abortion and nervous system of the fetus. Deficiency: Ocular anomalies
Vitamin D mcg	15	Excess: Hypercalcemia Deficiency: Reduced calcification and bone density, hypocalcaemia, neonatal tetany, and maternal osteomalacia
Vitamin C mg	120	Deficiency: Preterm labor and low birthweight
Vitamin E mg	19	Excess: Miscarriage
Thiamine (mg)	1.4	Not known
Riboflavin (mg)	3.4	Not known
Niacin (mg)	>20	Not known
B6 (mg)	2	Deficiency: Insufficient Apgar score
B 12 (μg)	2.8	Deficiency: Prematurity, CNS alterations, maternal megaloblastic anemia
Biotin (μg)	35	Not known
Folate (mcg)	600	Deficiency: Neural tube alterations, low birthweight

Modified por Food and Nutrition Board. RDA 2002. Food and Nutrition Board. National Academy of Sciences. National Research Council. Recommended dietary allowances. 11th ed. Washington, 2002

Nutrition with kidney transplantation

Pregnancy is not recommended with less than 2 years of transplantation, with proteinuria or with creatinine >1.5 mg/dL.

Strict glycemic control (corticosteroid effect) and maintenance of adequate weight. Eat at regular times, without sucrose and with frequent snacks [21] (Table 4.6).

Table 4.6 Daily nutritional recommendations for pregnant women with kidney transplantation [9,15,16].

Nutrient	Recommendation
Kilocalories	Individualized: Basal energy expenditure × physical activity + 300 kcal/day in 2nd–3rd trimester or 30 kcal/pregestational weight + 300 kcal/day Objective: To maintain maternal weight gain according to normal curves
Proteins	0.8–1 g/kg pregestational + 10 g/day according to metabolic need. Objective: To maintain serum urea levels <50 mg/dL, albumin >3.5 g/dL, creatinine <1.5 mg/dL
Na–K	Individualized 2–4 g/day
Phosphorus	Individualized: 1200 mg/day
Calcium	Individualized: 1200 mg/day

Nutritional therapy

During the first trimester of pregnancy, there is no effect on nutritional status unless there is previous malnutrition, folic acid deficiency can occur that can lead to congenital anomalies and neural tube deficiency.

During the second and third trimesters, nutritional status strongly influences fetal weight and prognosis.

A weight gain <6 kg means that there will be a greater consumption of maternal reserves, which can cause a low weight product with CNS alterations [21].

Indications for nutritional support

When serum albumin decreases by more than 1 g/L in the first trimester.

When there is low pregestational weight, weight loss, hyperemesis gravidarum that does not respond to nutritional supplement.

In the last trimester when weight gain does not exceed 6 kg and does not respond to nutritional advice and supplements.

Preserve lean body mass, maintain nutritional status, avoid other metabolic disorders, improve wound healing, support immune function, and reduce mortality [18].

The nutritional goals of enteral nutrition should be:

Weight gain according to prepregnancy weight status.

Maintain a positive nitrogen balance given by stabilization or increase in albumin.

Ensure adequate vitamin and mineral supplementation [22].

Enteral nutrition

Always consider this route before parenteral nutrition, it is the preferred option of nutritional supplementation. It has fewer potential metabolic and septic complications. Enteral feeding also takes advantage of the buffering effect of the gastrointestinal tract.

Parenteral nutrition

Maybe a safe way for patients who cannot tolerate oral or enteral nutrient administration.

Total parenteral nutrition (TPN) should be considered if enteral feeding is not sufficient or if the patient has gastrointestinal symptoms so severe that enteral feeding is not possible.

Both manage to improve nutritional parameters, if they reach the dietary goals of protein and energy, so the parenteral route is not considered superior to the oral route

[23,24]. 40%–60% carbohydrates, 20% amino acids, and 20%–40% fat are recommended. A water-soluble vitamin, including folic acid, should be provided. Control potassium, phosphorus, and magnesium that may decrease with treatment [25].

References

[1] Winfield S, Davison JM. Pregnancy and chronic kidney disease. In: Arici M, editor. Management of chronic kidney disease: a clinician's guide. 1st ed. Berlin, Heidelberg: Springer; 2014. p. 363–79.

[2] Alkhunaizi A, Melamed N, Hladunewich MA. Pregnancy in advanced chronic kidney disease and end-stage renal disease. Curr Opin Nephrol Hypertens 2015;24:252–9.

[3] Academy of Nutrition and Dietetics. EAL. Chronic Kidney Disease. Evidence-based nutrition practice guideline. U S Jpn Outlook: Academy of Nutrition and Dietetics; 2015 2010.

[4] Detsky AS, McLaughlin JR, Baker JP, Johnston N, Whittaker S, Mendelson RA, Jeejeebhoy KN. What is subjective global assessment of nutritional status? JPEN J Parenter Enteral Nutr 1987;11(1): 8–13. https://doi.org/10.1177/014860718701100108.

[5] Xia YA, Healy A, Kruger R. Developing and validating a renal nutrition screening tool to effectively identify undernutrition risk among renal inpatients. J Ren Nutr 2016;26:299–307.

[6] Rasmussen KM, Yaktine AL. Weight gain during pregnancy: reexamining the guidelines, 1. Nutrition. National U.S.:Academies Press; 2009. p. 250.

[7] Ismail LC, Bishop DC, Pang R, Ohuma EO, Kac G, Abrams B, et al. Gestational weight gain standards based on women enrolled in the Fetal Growth Longitudinal Study of the INTERGROWTH-21st project: a prospective longitudinal cohort study. BMJ 2016;352:1–10.

[8] Avesani CM, Kamimura MA, Cuppari L. Energy expenditure in chronic kidney disease patients. J Ren Nutr 2011;21:27–30. https://linkinghub.elsevier.com/retrieve/pii/S1051227610002888.

[9] Ikizler TA, Burrowes JD, Byham-Gray LD, et al. KDOQI clinical practice guideline for nutrition in CKD: 2020 update. Am J Kidney Dis 2020;76(3):S1–107. https://doi.org/10.1053/j.ajkd.2020.05.006. Suppl 1.

[10] Mousa A, Naqash A, Lim S. Macronutrient and micronutrient intake during pregnancy: an overview of recent evidence. Nutrients 2019;11(2).

[11] Garner C. Nutrition in pregnancy: dietary requirements and supplements. UpToDate 2022.

[12] Carrero JJ, Thomas F, Nagy K, et al. Global prevalence of protein-energy wasting in kidney disease: a meta-analysis of contemporary observational studies from the International Society of Renal Nutrition and Metabolism. J Ren Nutr 2018;28(6):380–92. https://doi.org/10.1053/j.jrn.2018.08.006.

[13] Ikizler TA, Cano NJ, Franch H, et al. Prevention and treatment of protein energy wasting in chronic kidney disease patients: a consensus statement by the International Society of Renal Nutrition and Metabolism. Kidney Int 2013;84:1096–107.

[14] Brookhyser J, Wiggins K. Medical nutrition therapy in pregnancy and kidney disease. Adv Ren Replace Ther 1998;5(No 1 (January):53–63.

[15] KDIGO clinical practice guideline for the evaluation and management of chronic kidney disease. Kidney Int 2013 Suppl;3:1–150.

[16] García MD, Escribano S, Manso de Real P, Sanchez Tocino ML, Arenas Jiménez MD. Qué novedades aportan en la práctica clínica las guías KDOQI de nutrición después de 20 años? NefroPlus 2022;14(1):1–10.

[17] Forney L, Doyle M. Standards of nutritional care for pregnant women. Seattle-King County Department of Public Health, Nutrition Services; 1991.

[18] Hou S. Pregnancy in women on haemodialysis and peritoneal dialysis. Baillieres Clin Obstest Gynecol 1994;8:481–500.

[19] Mahan LK, Escott-Stump S. Krause's food, nutrition and diet therapy. ed 14. Philadelphia, PA: Saunders; 2014.

[20] Food and Nutrition Board. National Academy of Sciences, National Research Council. Recommended dietary allowances. 11th ed. 2002. Washington.
[21] Wasserstein AG. L-carnitine supplementation in dialysis: treatment in quest of disease. Semin Dial. 2013;26(1):11–5. https://doi.org/10.1111/sdi.12041.
[22] Davison J. Dialysis, transplantation and pregnancy. Am J Kidney Dis 1991;17:127–32.
[23] Mac BM. Nutritional medicine: a case management approach. Philadelphia, PA: Saunders; 1989. p. 113–9.
[24] Compher C, Bingham A, Mc Call M, Patel J. Guidelines for the provision of nutrition support therapy in the adult critically ill patient: The American Society for Parenteral and Enteral Nutrition. JPEN J Parenter Enteral Nutr 2022;46(1):12–41. https://doi.org/10.1002/jpen.2267.
[25] Rombeau J, Richardson R. Clinical nutrition enteral and tube feeding. ed 3. Philadelphia, PA: Saunders; 1997. p. 439–57.

CHAPTER 5

Fluid management in critically ill pregnant patient

Kamal Lashkari
Department of Critical Care, Bombay Hospital, Indore, Madhya Pradesh, India

Introduction

In critically ill pregnant patients, maintaining appropriate fluid balance is of paramount importance to support organ perfusion and preserve uteroplacental flow. Pregnancy causes significant physiological adaptations that affect fluid management, including an increase in circulating volume and a decrease in systemic vascular resistance. As pregnancy progresses, the gravid uterus can mechanically obstruct the aorta and vena cava while supine, which can further complicate fluid management. Colloid oncotic pressure (COP) is another critical variable affected by pregnancy. Plasma and interstitial COP decrease throughout gestation, and there is a corresponding increase in capillary hydrostatic pressure. An increase in hydrostatic pressure or decrease in plasma COP may favor edema formation in late pregnancy, potentially leading to acute pulmonary edema, which is a life-threatening condition. Certain disease states, such as preeclampsia, can also affect overall fluid physiology in pregnancy, posing additional challenges in effective fluid therapy [1–4].

The appropriate type of fluid for critically ill pregnant patients is not clear, and there is a lack of high-quality evidence to guide clinical decision-making. Isotonic crystalloids are generally considered safe for initial fluid therapy, while synthetic colloids and starches carry significant risks and should be avoided in critical illness. Although normal saline is the most commonly used fluid in critical care, there is growing concern regarding saline-induced hyperchloremic metabolic acidosis and renal injury. Balanced salt solution is being studied to establish its role in fluid therapy. Albumin solutions may be safe, but there is limited evidence for their widespread utility [1,3,4].

The assessment of volume status in critically ill pregnant patients is a dynamic and challenging process that requires frequent monitoring and high clinical acumen. To optimize hemodynamic management and improve patient outcomes, the principle of "fluid stewardship" has been proposed. It is guided by the four rights of medication safety: right patient, right drug, right route, and right dose. A standard approach to the assessment of fluid needs and prescription of intravenous (IV) fluid therapy is necessary as patient-specific considerations for fluid therapy evolve across the continuum of critical illness. Future research should focus on areas guided by the principle of fluid stewardship,

The Kidney of the Critically Ill Pregnant Woman
ISBN 978-0-443-21473-8,
https://doi.org/10.1016/B978-0-443-21473-8.00016-1

emphasizing the need for high-quality evidence to guide fluid therapy in critically ill pregnant patients, and addressing the challenges posed by varying healthcare settings with different resource and skill availability [1,5,6].

Physiological changes in pregnancy

During pregnancy, significant physiological changes occur to accommodate the growing fetus. Blood volume gradually increases up to 1.5 times that of nonpregnant individuals by the end of 32 weeks, with a parallel increase in red cell mass that is proportionately less, resulting in a decrease in hematocrit. The overall reduction in plasma oncotic pressure due to hemodilution can make the patient susceptible to pulmonary edema, despite a reduction in pulmonary capillary wedge pressure. Cardiovascular changes include a gradual increase in cardiac output in response to the increase in blood volume, which is approximately 40% by the end of the second trimester. This increase is due to both stroke volume and heart rate. Additionally, baseline blood pressure and systemic vascular resistance decrease, with a greater reduction in diastolic blood pressure than systolic. These adaptive changes are more pronounced in multiple pregnancies. The position of the mother is critical, as compression of the inferior vena cava by the gravid uterus in the supine position may lead to a 25% reduction in cardiac output. Renal blood flow and glomerular filtration rate also increase significantly. The increased flow leads to dilation of the calyx and ureter, and the increased glomerular capillary permeability can cause glycosuria and proteinuria, contributing to an increased risk of bacteriuria. Respiratory system adaptation includes a 40% increase in minute ventilation, primarily due to an increase in tidal volume rather than respiratory rate, and a reduction in functional residual capacity due to diaphragmatic elevation [7,8].

Common indications of fluid therapy in critically ill pregnant

Traditionally, fluid therapy was categorized into three types: resuscitation, replacement, and maintenance therapy. However, with mounting evidence supporting the hypothesis that fluid overload can have negative impacts on patient outcomes, newer fluid therapy models have been emerging. The ROSE model has been introduced for sepsis, representing resuscitation, optimization, stabilization, and evacuation. Furthermore, fluid stewardship has been suggested, emphasizing the 4D concept—drug, dose, duration, and deescalation.

In critically ill pregnant patient, fluids can be given for a number of reasons, including treating hypotension, replacing lost fluids, and improving cardiac output and tissue perfusion. Causes of significant bleeding during pregnancy may include uterine atony, retained placenta, trauma, placenta previa, or abruption placentae. Fluid therapy may also be used to address conditions such as pregnancy-induced hypertension and infections leading to sepsis or septic shock. However, the decision to administer fluids is complex and depends on a number of factors, including whether the underlying condition is responsive to fluids and the potential risks associated with administering fluids in a given clinical context. Additionally, in later stages of pregnancy, hypotension may occur due to the pressure of the gravid uterus on the inferior vena cava, which can be alleviated by lateral maternal positioning or displacement of the uterus.

Electrolyte correction, acid—base disorder correction, glucose, or artificial nutrition provision, and IV medication delivery are some other reasons for fluid therapy. Effective fluid administration in critically ill pregnant patients requires a drug-like approach, involving determining the indication, prescribing the appropriate dose, and assessing the response [1,4,6,9,10].

Fluid compartments

On average, females have a total body water percentage of approximately 50%. However, during pregnancy, there are significant physiological changes that occur, such as a disproportionate increase in total blood volume and red cell mass. Consequently, the total body water can increase to around 60%—65% of the female's weight. The intracellular fluid constitutes two-thirds of the total body water, while the remaining one-third is extracellular fluid. Further extracellular fluid is subdivided in interstitial and intravascular compartments. 75% of extracellular fluid is interstitial and the remaining is intravascular [3,11].

Common fluids

Crystalloids

Crystalloid fluids are commonly used to resuscitate patients suffering from hypovolemia, hemorrhage, sepsis, and dehydration. In addition, they can be used to deliver IV medications, provide maintenance fluids to patients with limited enteral nutrition, manage blood pressure, and increase diuresis to prevent end-organ damage from nephrotoxic drugs or toxins. These fluids contain physiologically active solutes, with sodium chloride being the primary cation and determinant of osmotic pressure in the extracellular fluid. The primary effect of crystalloid fluids is to expand the interstitial space, with only 20% remaining in the intravascular compartment after 30 minutes.

Normal saline

Properties (pH 4.2, Na 154 mEq/L, Cl-154 mEq/L, 308 mOsm/kg).

Normal saline at 0.9% is commonly used for acute volume expansion due to its sodium content, which is a major cation and determinant of extracellular fluid osmotic pressure. Its osmolality is similar to that of plasma, which prevents significant osmotic movement of water. However, its high chloride content raises concerns about hyperchloremic metabolic acidosis with large volume infusions. Large volume administration can also lead to hypochloremia-induced renal afferent arteriole constriction and reduced GFR, as demonstrated by some studies.

Several high-quality trials have investigated hyperchloremia-induced kidney dysfunction in ICU patients, but the results have been inconsistent. Normal saline has been compared with balanced salt solutions, such as Ringer lactate (RL) or PlasmaLyte A, in several studies, including the SALT-ED trial in the emergency department and the SMART trial in the ICU. These studies revealed that normal saline may be harmful, as it is associated with more Major Adverse Kidney Events within 30 days (MAKE30 events), which defined as composite of death, requirement for new renal replacement therapy, or persistent renal dysfunction. However, recent trials, including BASICS and Plus trial, have failed to demonstrate harm caused by normal saline in ICU patients. Additionally, the BASICS trial demonstrated the safety of normal saline in traumatic brain injury patients. Although the variable results of randomized controlled trials, combined with the high cost of balanced salt solutions, make it unlikely to be the universal fluid of choice in the ICU, it can be preferred in critically ill patients with altered acid—base balance, acute kidney injury (AKI) as those usually have narrow reserve of buffers [1,3,12—16].

Ringer lactate

Properties; pH 6.5, Na 130 mEq/L, Cl-109 mEq/L, K 4 mEq/L, Lactate 28 mEq/L, Ca3.0 mEq/L, 273 mOsm/kg.

RL is a balanced electrolyte solution that can be used as an alternative to normal saline in hypovolemia or shock. RL contains sodium, potassium, and chloride ions in the same concentrations as found in plasma. It also includes lactate as a buffer, which is metabolized to bicarbonate in the liver. Major advantage of RL over normal saline is it does not produce hyperchloremic metabolic acidosis. However, caution should be exercised when using RL in patients with hepatic and renal dysfunction, as lactate metabolism can be impaired in these individuals. Additionally, RL should be avoided in patients with diabetic ketoacidosis (DKA) and hyperosmolar hyperglycemic state (HHS) as it can worsen hyperglycemia due to the presence of lactate [1,3,17,18].

PlasmaLyte

Properties; pH 7.4, Na 140 mEq/L, Cl 98 mEq/L, K 5 mEq/L, Mg 3.0 mEq/L, 295 mOsm/kg. Buffer acetate 27 mEq/L, gluconate 23 mEq/L.

PlasmaLyte is a highly buffered fluid with which is most equivalent to plasma in terms of pH, tonicity. The fluid composition addresses the concrete of hypochloremia and metabolic acidosis. However, usual cost of commercially available fluids is usually very high [3].

Hypotonic saline 0.45%

Hypotonic fluids, such as one-half normal saline, 0.45%, are not effective for acute volume expansion as they quickly lose their volumes to extravascular space due to hypotonicity. They are not preferred resuscitation fluid. However, this fluid is been considered in maintaining IV fluid in DKA to avoid hyperchloremic acidosis [3,18].

Hypertonic sodium-containing solutions

Hypertonic sodium-containing solutions are recommended for patients who require sodium supplementation but cannot tolerate large fluid volumes due to conditions such as neurologic injury or distributive shock. In neurologic injury, hypertonic solutions help reduce elevated intracranial pressure by removing fluid from the intracranial extravascular compartment. In distributive shock, a hypertonic solution may be preferred as it requires smaller total volumes and can mobilize edema while increasing circulating blood volume. However, caution should be exercised in patients with compromised cardiac or renal function. Additionally, it requires close monitoring of electrolyte and osmolality. Although hypertonic solutions have not been associated with complications in trauma resuscitation, they may not be suitable for shock states where total body fluid is decreased, such as hemorrhagic or hypovolemic shock. Further studies are necessary to fully understand the physiological implications of hypertonic fluid resuscitation strategies [3].

Dextrose based fluids

Properties; 5%DW, 5 g/dl, 278 mOsm/kg.

Dextrose-based fluids are not suitable for volume expansion due to their tendency to rapidly enter the intracellular compartment. Moreover, they should be avoided in patients with increased intracranial pressure. They are only used in specific situations such as hypoglycemia or as part of parenteral nutrition, as they can alter serum osmolality and cause metabolic imbalances [3].

Colloids

Colloids and crystalloids are options for fluid resuscitation, with colloid use based on rationale that it expands vascular volume and decreases fluid leakage. The duration of plasma expansion may last from 2 to 36 hours. However, recent trials show that starch-based colloids increase renal dysfunction and mortality risk. Guidelines do not recommend the routine use of colloids. Individual factors, such as fluid balance and medical conditions, should be considered in selecting fluids [3,19,20].

Human serum albumin

Albumin is a crucial transport protein synthesized in the liver that serves as the primary oncotic protein of plasma. It constitutes about 80% of the colloid oncotic pressure of plasma and is commercially available in several preparations, including 4%, 5%, 20%, and 25% solutions in isotonic saline. The 5% solution has an equivalent colloid oncotic pressure (COP) to plasma, whereas the 25% solution has a significantly higher COP. Infusing 100 mL of the 25% solution can expand plasma volume up to 500 mL, and its effects can last for more than 24 hours. Despite its widespread use, studies such as SAFE and ALBIOS have failed to establish any mortality benefit, and targeting a serum albumin level of 30 g/L or higher does not enhance survival. Moreover, using albumin for routine fluid resuscitation may not be cost-effective. While the Surviving Sepsis guideline weakly recommends its use in individuals who have received large quantities of crystalloid, caution should be exercised when using large volumes to avoid dilutional coagulopathy. Furthermore, it is worth noting that albumin eventually passes into the interstitial space and may cause delayed pulmonary edema in patients at risk [3,21,22].

Hydroxyethyl starch

Hetastarch or hydroxyethyl starch (HES) is a synthetic colloid used as a less expensive alternative to albumin. Its concentration influences the volume effect, with a 6% solution being isotonic and a 10% solution being hyperoncotic. HES has a longer half-life than albumin, with 50% of the osmotic effect persisting for 24 hours. While HES products decrease circulating plasma concentrations of coagulation factor VIII and von Willebrand factor and impair platelet activity, coagulopathies are less common than albumin. However, numerous studies have demonstrated nephrotoxic effects of hyperoncotic HES solutions, and the use of HES should be generally discouraged, including in the critical care and perioperative settings. Neither form of HES solution should probably be used in patients with trauma or any form of brain injury.

Several randomized controlled trials have compared the use of HES and crystalloids in critically ill patients. The CRISTAL trial found no difference in 28-day mortality between the two groups, but 90-day mortality was lower in the colloid group. The CHEST trial found no difference in 90-day mortality, but HES was associated with more adverse events. The 6S trial found an increased risk of death and more renal replacement therapy in the HES group, while the VISEP trial found an increased rate of AKI and renal replacement therapy in the HES group [3,23,24].

Dextran

Dextrans are a type of synthetic colloid that are derived from sugar beet juice and are polysaccharides. Two commonly used dextrans are dextran40 and dextran70, with mean molecular weights of 40,000 and 70,000, respectively. Dextran40 is available as a 10% solution with a COP of 40 mm Hg, and its acute volume expansion is about twice the infused volume. However, over 50% is cleared after 6 hours. Dextrans are effective artificial colloids, but their use is limited due to significant side effects. Dextrans are mainly used to improve microcirculatory flow in microsurgical reimplantations and in extracorporeal circulation during cardiopulmonary bypass. Dextrans cause anaphylactic reactions, coagulation abnormalities, and interfere with cross-match. They are also associated with renal failure and should not be used for volume resuscitation [3,23,24].

Gelatin

Gelatin is a substance that is made by breaking down the collagen of bovine or porcine sources. There are two common types of gelatin available, which are succinylated gelatin and urea-linked gelatin-polygeline, and they come in 500 mL plastic containers. Gelatin has a low molecular weight and it only expands the plasma volume for a short time before being metabolized and excreted by the kidneys. The use of gelatin-based fluids can increase the risk of AKI due to its accumulation in the reticuloendothelial system. Gelatin-based fluids can also cause anaphylaxis, which can be life-threatening. Gelatin solutions have similar toxicities to HES solutions, and both substances can cause capillary dilation, epithelial cell damage, and increased cell death. Gelatin is not recommended as a resuscitation fluid due to the risks of anaphylaxis, bleeding, AKI, and mortality [23–25].

Hemodynamic monitoring

Hemodynamic monitoring is crucial for critically ill pregnant patients and can be categorized into clinical monitoring, biochemical monitoring, and advanced hemodynamic monitoring. Clinical monitoring involves assessing clinical parameters, while biochemical monitoring includes laboratory markers of tissue perfusion. Advanced hemodynamic monitoring mainly focuses on advanced parameters such as cardiac index, cardiac output, systemic vascular resistance, and extravascular lung water. Monitoring systems can be classified as invasive, minimally invasive, or noninvasive. Invasive monitoring systems, such as the pulmonary artery catheter, have been limited due to risks and complications. Minimally invasive systems are based on arterial pressure waveform analysis and include PiCCO, LiDCO, and Flotrac. Noninvasive monitoring systems include transthoracic echocardiography, transesophageal echocardiography, bioimpedance, and bioreactance. Dynamic hemodynamic monitoring, based on heart–lung interaction, involves analyzing arterial waveform to derive pulse pressure variation and stroke volume variation. Using passive leg raising (PLR) to induce a gravitational transfer of blood and increase cardiac preload is particularly helpful in spontaneously breathing patients. PLR is a rapid, noninvasive, and reversible method to evaluate fluid responsiveness. There is some evidence indicating that the hemodynamic response is comparable between nonpregnant individuals and pregnant women at 22–24 weeks. Nonetheless, its effectiveness during the third trimester of pregnancy has not been verified [26–29].

Specific maternal conditions

Pregnancy-induced hypertension

The predominant gestational ailment impacting fluid equilibrium is preeclampsia. The fluid management aspect is especially complex, as this subset of patients can present two extremes that guide treatment protocols regarding fluid regulation: pulmonary edema and renal dysfunction. These patients are highly susceptible to acute pulmonary edema, which is a significant cause of mortality [27,30,31].

Development of pulmonary edema in patients with pregnancy-induced hypertension is attributed to several common factors such as decreased oncotic pressure, heightened capillary permeability, edema resolution during the purpureal stage, augmented afterload due to vasoconstriction, and cardiac pathologies such as cardiomyopathy and valvular heart diseases. Additionally, excessive IV fluid administration can also be a contributing factor [30,32].

IV fluid therapy is utilized for several indications in patients with preeclampsia. Resuscitation is one such indication as these patients are at an increased risk of severe hypovolemia and shock following blood loss from conditions like placental abruption, operative loss, and postpartum hemorrhage (PPH). The common principle to manage this condition involves resuscitation with crystalloids, transfusion of blood products, and avoiding the triad of coagulopathy, metabolic acidosis, and hypothermia [30].

Replacement therapy is given based on deficit estimation, but AKI may complicate the replacement due to the risk of fluid overload and pulmonary edema. Advanced hemodynamic monitoring can prevent such complications. Correcting anemia and coagulopathy warrants replacement with suitable blood products [27,30,31].

Maintenance fluids are necessary for patients who cannot take food orally and act as carriers for drugs such as magnesium therapy. Usually, 60–80 mL per hour of crystalloid, including all drug infusions, should suffice without increasing the risk of pulmonary edema. An accepted minimum urine output of 0.5 mL/kg/hour or 2 mL/kg over 4 hours should be maintained, and intermittent fluid challenges should be given cautiously to maintain this output [30,32,33].

Proper fluid preloading should be performed before regional anesthesia with 300 mL of crystalloid fluid. Although many experts prefer colloidal solutions for volume preloading, there is limited data regarding their safety in critically ill pregnant patients since these subgroups are usually excluded from randomized control trials [27,30,31,34,35].

As severe preeclampsia patients represent severely vasoconstricted vascular system combined with increased vascular permeability. These patients see sudden drop of blood pressure on vasodilatation and in response to dose of antihypertensive drugs like hydralazine. Due to such vulnerability to rapid development of hypotension and pulmonary edema, they often require invasive hemodynamic monitoring. However noninvasive strategies including bed side 2D echo can give enough clinical information, performing a fluid challenge and passive leg raise can be difficult due to aortocaval compression by gravid uterus [4,36,37].

Managing renal failure

Peripartum/postpartum is period with maximum fluid shifts. Patient may suffer oliguria and sometimes AKI. There is no specific fluid strategy proven to reverse the AKI. Usually, maintenance of fluid should not be manipulated more than 1 mL/kg/hour to target urine output of 100 every four hourly. Invasive hemodynamic monitoring (or minimally invasive cardiac output monitoring) should be considered to guide fluid therapy [4,30,36].

Obstetrics hemorrhage

Obstetric hemorrhage is one of important causes of maternal mortality worldwide. There are several definitions suggested for PPH. According to the American College of Obstetricians and Gynecologists (2017), it is defined as loss of more than 1 L blood irrespective of route of delivery or any blood loss that causes hemodynamic instability [38].

Physiological changes of pregnancy like tachycardia, reduced SVR, and high pulse pressure can obscure the clinically significant bleeding. Physicians are likelier to underestimate the blood loss. Therefore, blood product transfusion may be delayed. Moreover, blood loss more likely to risk oxygen carrying capacity in already physiological anemia state [4]. The common causes of PPH can be remembered with the mnemonic 4 T, that is, tone (uterine atony), trauma, tissue (retained products), and thrombin. Uterine atony usually accounts for more than 70% of bleeding [39]. Management should focus on immediate control of hemorrhage, replacement of volume, and avoiding or addressing the consequences like including hypothermia, coagulopathy, metabolic acidosis, and multiorgan dysfunction [4].

Initial management according to FIGO guidelines 2022, bundle care includes the use of uterotonic drugs, isotonic crystalloids, tranexamic acid, and uterine massage [38].

Crystalloids are fluid of choice for resuscitation. Potential harm of normal saline induced kidney injury, metabolic acidosis should be kept in mind in subjects who are receiving larger volumes and are at greater risk of developing such complications. Balanced salt solutions can be considered in such patients. Restrictive fluid strategy which targets mean arterial pressure 55–60 mmHg has been advocated especially once fetus is delivered, though evidence is lacking to support same. The hypothesis behind restrictive strategy is to avoid the dislodgement of clot with higher blood pressure and worsening of coagulopathy due to hemodilution. Hence blood products should be arranged and transfused earliest possible. The optimal ratio of packed red blood cells (RBCs), plasma, and platelets is still under investigations and ideally should be individualized [38,40–42].

When initial measure for controlling PPH fails, bundle for refractory PPH has been proposed which includes compressive measures (aortic compression or bimanual uterine compression), intrauterine balloon tamponade, and nonpneumatic antishock garment [38].

Trauma in pregnancy

Trauma is a major cause of maternal mortality during pregnancy, with injuries accounting for 20% of maternal deaths. Therefore, it is recommended that all injured women of child-bearing age should be considered pregnant until proven otherwise. Initial assessment of an acutely injured pregnant woman should include securing the airway, ensuring breathing, and maintaining circulation. Standard trauma protocols for administration of fluids and blood products during resuscitation should be followed, with the insertion of two large bore needles recommended for initial rapid crystalloid infusion and intravascular volume expansion. However, vasopressor use should be delayed if possible due to their potential to decrease placental perfusion in cases of maternal hypotension. Crystalloids are fluid of choice for initial resuscitation, however should be restricted in favor of blood products, as they can exacerbate bleeding and reduce oxygen delivery. Displacement of the uterus off of the inferior vena cava and abdominal aorta can improve uterine perfusion and be achieved through the left lateral position, manual displacement of the uterus, or the use of a backboard for lateral tilt. In cases of acute blood loss, O-negative blood should be transfused until type-specific or cross-matched donor blood becomes available to avoid Rh sensitization in Rh-negative women. The Kleihauer—Betke blood test should be sent to calculate dosing of anti-D immunoglobulin for Rh-negative mothers exposed to Rh-positive blood and to assist in fetal prognostication after major trauma. Tocodynamometric monitoring should be performed if placental abruption is suspected.

Maternal cardiac arrest

When managing maternal cardiac arrest, adult life support principles should be followed, but modifications should be made after 20 weeks of gestation. Uterine displacement should be performed immediately, and manual displacement is preferred over left lateral tilt to allow for more effective cardiac compression. Resuscitative hysterotomy should be performed within 4 minutes of arrest to relieve aortocaval pressure, improve respiratory dynamics, and provide placental autotransfusion to optimize chest compressions, oxygen delivery, and cardiac output simultaneously. It also provides the best chance of fetal survival if a viable gestational age (24 weeks) has been reached [43—47].

Maternal sepsis

Sepsis is characterized by life-threatening organ dysfunction resulting from an unregulated host response to infection. The Society of Obstetric Medicine of Australia and

New Zealand (SOMANZ) recommends the use of omqSOFA and omSOFA scores, which take into account the physiological changes that occur during pregnancy, for assessing sepsis in pregnant women. IV fluid resuscitation is the first line of treatment for septic shock. The RCOG recommended initial fluid resuscitation rate is 20 mL/kg, as opposed to the standard 30 mL/kg, due to an increased risk of pulmonary edema caused by decreased COP. Crystalloids are preferred over colloids as they are associated with renal toxicity. Although albumin is considered safe for use in septic shock, there is no clear benefit over crystalloid, and the increased cost makes crystalloid the preferred initial resuscitation fluid in sepsis cases. Prompt administration of antibiotics is mandatory as delay in administering antibiotics can increase maternal mortality by 8% per hour. Norepinephrine is the first-choice vasopressor for maternal sepsis, and the mean arterial pressure should be individualized based on overall organ perfusion. Further fluid therapy should be guided by dynamic parameters of hemodynamic monitoring [48—55].

Heart failure in pregnancy

Heart failure during pregnancy has various causes including cardiomyopathies, coronary artery disease, valvular heart disease, adult congenital heart disease, arrhythmias, and pulmonary hypertension. Risk factors for cardiomyopathy include viral myocarditis, autoimmune factors, alcohol, and drug abuse. Peripartum cardiomyopathy is an idiopathic condition with heart failure near delivery. Management is supportive, and recovery of left ventricular function is common. Valvular disease can result from rheumatic heart disease or congenital valvular heart disease [56—59].

The treatment of heart failure during pregnancy involves diuretics as first-line therapy, with nitroglycerine used for acutely ill pregnant women with hypertension and systolic failure. Beta-blockers and spironolactone can be used for systolic dysfunction, and bromocriptine may be used to suppress lactation in peripartum cardiomyopathy. Arrhythmias may be managed with drugs or implantable defibrillators. Inotropic support and assist devices may be necessary for severe ventricular dysfunction. Obstetric management depends on the contribution of pregnancy physiology to heart failure. Pulmonary edema developing in preeclamptic mothers is an indication for the termination of pregnancy due to its high fatality rate and clear association with maternal mortality [58—61].

Pulmonary embolism

During pregnancy, the risk of thrombotic events increases due to the hypercoagulable state resulting from elevated levels of clotting factors. Diagnosis of pulmonary embolism (PE) in pregnant women is challenging due to overlapping clinical symptoms with physiological changes in pregnancy. Multidetector-row CTA is the criterion standard for diagnosing PE in pregnant women, and it can be safely performed in all trimesters.

Treatment of PE is based on risk stratification, and low molecular weight heparin (LMWH) is the preferred anticoagulant for pregnant patients. Patients with high-risk PE should receive prompt IV anticoagulation with unfractionated heparin (UFH). Obstetric patients diagnosed with obstructive shock should receive thrombolysis, although those in the early postpartum period are at higher risk for complications [62–69].

References

[1] Hawkins WA, Smith SE, Newsome AS, Carr JR, Bland CM, Branan TN. Fluid stewardship during critical illness: a call to action. J Pharm Pract 2020;33(6):863–73. https://doi.org/10.1177/0897190019853979.

[2] Dennis AT, Solnord CB. Acute pulmonary oedema in pregnant women. Anaesthesia 2012;67:646–59. https://doi.org/10.1111/j.1365-2044.2012.07055.x.

[3] Graves CR. Fluid and electrolyte therapy in the critically ill obstetric patient. In: Foley MR, Lee TR, Archbold TR, editors. Obstetric intensive care manual. 4th ed. Wiley-Blackwell; 2021 [Chapter 25].

[4] Crozier T. Fluid management in the complicated obstetric patient. In: Lapinsky S, editor. Principles and practice of obstetric high-dependency and critical care (obstetrics module). The continuous textbook of women's medicine series, 9. Wiley-Blackwell; 2021. https://doi.org/10.3843/GLOWM.413243.

[5] Messmer AS, Zingg C, Müller M, Gerber JL, Schefold JC, Pfortmueller CA. Fluid overload and mortality in adult critical care patients – a systematic review and meta-analysis of observational studies. Crit Care Med 2020;48(2):186–93. https://doi.org/10.1097/CCM.0000000000004139.

[6] Malbrain MLNG, Van Regenmortel N, Saugel B, De Tavernier B, Van Gaal PJ, Joannes-Boyau O, Teboul JL, Rice TW, Mythen M, Monnet X. Principles of fluid management and stewardship in septic shock: it is time to consider the four D's and the four phases of fluid therapy. Ann Intensive Care 2018;8(1):66. https://doi.org/10.1186/s13613-018-0410-9.

[7] Soma-Pillay P, Nelson-Piercy C, Tolppanen H, Mebazaa A. Physiological changes in pregnancy. Cardiovasc J Afr 2016;27(2):89–94. https://doi.org/10.5830/CVJA-2016-021.

[8] Vinturache A, Khalil A. Maternal physiological changes in pregnancy. In: Khalil A, editor. Fetal development and maternal adaptation (obstetrics module). The continuous textbook of women's medicine series, 4. Wiley-Blackwell; 2021. https://doi.org/10.1002/9781119096517.

[9] Malbrain MLNG, Marik PE, Witters I, Cordemans C, Kirkpatrick AW, Roberts DJ, Van Regenmortel N. Fluid overload, de-resuscitation, and outcomes in critically ill or injured patients: a systematic review with suggestions for clinical practice. Anaesthesiol Intensive Ther 2014;46(5):361–80. https://doi.org/10.5603/AIT.2014.0060.

[10] Hoste EA, Maitland K, Brudney CS, Mehta R, Vincent JL, Yates D, Kellum JA, Mythen MG, Shaw AD, Adqi XII Investigators Group. Four phases of intravenous fluid therapy: a conceptual model. Br J Anaesth 2014;113(5):740–7. https://doi.org/10.1093/bja/aeu300.

[11] Tobias A, Ballard BD, Mohiuddin SS. Physiology, water balance. In: StatPearls [Internet]. Treasure Island (FL): StatPearls Publishing; 2023.

[12] Self WH, Semler MW, Wanderer JP, et al. Balanced crystalloids versus saline in noncritically ill adults. N Engl J Med 2018;378(9):819–28. https://doi.org/10.1056/NEJMoa1711584.

[13] Semler MW, Self WH, Wanderer JP, et al. Balanced crystalloids versus saline in critically ill adults. N Engl J Med 2018;378(9):829–39. https://doi.org/10.1056/NEJMoa1711584.

[14] Chang R, Holcomb JB. Choice of fluid therapy in the initial management of sepsis, severe sepsis, and septic shock. Shock 2016;46(1):17–26. https://doi.org/10.1097/SHK.0000000000000577.

[15] Zampieri FG, Machado FR, Biondi RS, et al. Effect of intravenous fluid treatment with a balanced solution vs 0.9% saline solution on mortality in critically ill patients: the BaSICS randomized clinical trial. JAMA 2021;326(9):1–12. https://doi.org/10.1001/jama.2021.11684.

[16] Finfer S, Micallef S, Hammond N, Navarra L, Bellomo R, Billot L, Delaney A, et al. Balanced multi-electrolyte solution versus saline in critically ill adults. N Engl J Med 2018;378(9):829–39. https://doi.org/10.1056/NEJMoa1711586.

[17] Singh S, Kerndt CC, Davis D. Ringer's lactate. In: StatPearls [Internet]. Treasure Island (FL): StatPearls Publishing; 2023.

[18] Gershkovich B, English SW, Doyle MA, Menon K, McIntyre L. Choice of crystalloid fluid in the treatment of hyperglycemic emergencies: a systematic review protocol. Syst Rev 2019;8(1):228.

[19] Myburgh JA, Finfer S, Bellomo R, et al. Hydroxyethyl starch or saline for fluid resuscitation in intensive care. N Engl J Med 2012;366(24):2165–76. https://doi.org/10.1056/NEJMoa1209759.

[20] Perner A, Haase N, Guttormsen AB, et al. Hydroxyethyl starch 130/0.42 versus Ringer's acetate in severe sepsis. Crit Care 2012;16(3):R94. https://doi.org/10.1186/cc11358.

[21] Finfer S, Bellomo R, Boyce N, French J, Myburgh J, Norton R, et al. A comparison of albumin and saline for fluid resuscitation in the Intensive Care Unit. JAMA 2004;291(7):829–36.

[22] Caironi P, Tognoni G, Masson S, Fumagalli R, Pesenti A, Romero M, et al. Albumin replacement in patients with severe sepsis or septic shock. N Engl J Med 2014;370(15):1412–21.

[23] Mayerhöfer T, Shaw AD, Wiedermann CJ, Joannidis M. Fluids in the ICU: which is the right one? Nephrol Dial Transplant 2022;28:1–10. https://doi.org/10.1093/ndt/gfac279.

[24] Mitra S, Khandelwal P. Are all colloids same? How to select the right colloid? Indian J Anaesth 2009; 53(5):592–607.

[25] Finfer S, Myburgh J, Bellomo R. Intravenous fluid therapy in critically ill adults. Nat Rev Nephrol 2018;14:541–57. https://doi.org/10.1038/s41581-018-0022-6.

[26] Rojas-Suarez J, Borré D, Zapata R. Hemodynamic monitoring of the critically ill obstetric patient. In: Lapinsky S, editor. Principle and practice of obstetric high dependency and critical care. The continuous textbook of women's medicine series – obstetrics module, 9. University of Toronto, Canada: MedEdPORTAL Publications; 2021.

[27] Marik PE, Monnet X, Teboul JL. Hemodynamic parameters to guide fluid therapy. Ann Intensive Care 2011;1(1):1. https://doi.org/10.1186/2110-5820-1-1.

[28] Vårtun Å, Flo K, Acharya G. Effect of passive leg raising on systemic hemodynamics of pregnant women: a dynamic assessment of maternal cardiovascular function at 22–24 weeks of gestation. PLoS One 2014;9(4):e94629. https://doi.org/10.1371/journal.pone.0094629.

[29] Burlinson CEG, Sirounis D, Walley KR, et al. Sepsis in pregnancy and the puerperium. Int J Obstet Anesth 2018;36:96–107. https://doi.org/10.1016/j.ijoa.2018.04.010.

[30] Anthony J, Schoeman LK. Fluid management in pre-eclampsia. Obstet Med 2013;6(3):100–4. https://doi.org/10.1177/1753495X13486896.

[31] Agency for Clinical Innovation. Maternity – management of hypertensive disorders of pregnancy. Policy Directive PD2011_064. Patient Matters policy manual. Publication date 2011 Oct 12. Review date 2023 Sep 28. Functional group Clinical/Patient Services – Maternity, Pharmaceutical. Author branch contact: (02) 9464 4711. File number 11/5201. 2011.

[32] Krishnachetty B, Plaat F. Management of hypertensive disorders of pregnancy. Anaesthesia Tutorial of the Week 2014:304.

[33] NICE. Hypertension in pregnancy: diagnosis and management. NICE Guideline; 2019. Available at: www.nice.org.uk/guidance/ng133.

[34] Narkhede AM, Karnad DR. Preeclampsia and related problems. Indian J Crit Care Med 2021; 25(Suppl. 3):S261–6.

[35] Imarengiaye CO, Isesele TO. Intensive care management and outcome of women with hypertensive diseases of pregnancy. Niger Med J 2015;56(5):333–7. https://doi.org/10.4103/0300-1652.170389.

[36] Lam MTC, Dierking E. Intensive Care Unit issues in eclampsia and HELLP syndrome. Int J Crit Illn Inj Sci 2017;7(3):136–41. https://doi.org/10.4103/IJCIIS.IJCIIS_33_17.

[37] Dennis AT, Solnordal CB. Acute pulmonary oedema in pregnant women. First published: 15 March 2012. doi: 10.1111/j.1365-2044.2012.07055.

[38] Mavrides E, Allard S, Chandraharan E, Collins P, Green L, Hunt BJ, Riris S, Thomson AJ, on behalf of the Royal College of Obstetricians and Gynaecologists. Prevention and management of postpartum haemorrhage. BJOG 2016;124:e106–49.

[39] Anderson JM, Etches D. Prevention and management of postpartum hemorrhage. Am Fam Physician 2007;75(6):875–82.
[40] Smith CM, Borders AE, et al. Quantitative blood loss in obstetric hemorrhage. Committee Opinion No. 794. American College of Obstetricians and Gynecologists. Obstet Gynecol 2019;134(6): e150–7. https://doi.org/10.1097/AOG.0000000000003589.
[41] WHO recommendations for the prevention and treatment of postpartum haemorrhage. Geneva: World Health Organization; 2012. https://apps.who.int/iris/bitstream/handle/10665/75411/9789241548502_eng.pdf.
[42] Ickx BE. Fluid and blood transfusion management in obstetrics. Eur J Anaesthesiol 2010;27:1031–5. https://doi.org/10.1097/EJA.0b013e32833e7a9d.
[43] Jain V, Chari R, Maslovitz S, et al. Sogc CLINICAL practice guideline: guidelines for the management of a pregnant trauma patient. J Obstet Gynaecol Can 2015;37(6):553–71.
[44] Irving T, Menon R, Ciantar E. Trauma during pregnancy. BJA Educ 2021;21(1):10–9. https://doi.org/10.1016/j.bjae.2020.08.005.
[45] Garg N, Sharma A, Khanna P, Goel V. Trauma in pregnancy – a brief review. Trauma Emerg Care 2017;2(3):1–4. https://doi.org/10.15761/TEC.1000131.
[46] Murphy NJ, Quinlan JD. Trauma in pregnancy: assessment, management, and prevention. Am Fam Physician 2014;90(10):717–22.
[47] Mendez-Figueroa H, Dahlke JD, Vrees RA, Rouse DJ. Trauma in pregnancy: an updated systematic review. Am J Obstet Gynecol 2013;209(1):1–10. https://doi.org/10.1016/j.ajog.2013.01.021.
[48] Bowyer L, Robinson HL, Barrett H, et al. SOMANZ guidelines for the investigation and management sepsis in pregnancy. Aust N Z J Obstet Gynaecol 2017;57:540–51. https://doi.org/10.1111/ajo.12646.
[49] Ali A, Lamont RF, et al. Recent advances in the diagnosis and management of sepsis in pregnancy. F1000Research 2019;8:1546. https://doi.org/10.12688/f1000research.18736.1.
[50] Greer O, Shah NM, Johnson MR, et al. Maternal sepsis update: current management and controversies. Obstet Gynaecol 2020;22:45–55. https://doi.org/10.1111/tog.12623.
[51] Cheshire J, Dunlop C, Lissauer D, et al. The management of maternal sepsis. In: Kate Lightly D, editor. The continuous textbook of women's medicine series – obstetrics module. 15. Glob. Libr. Women's Med.; 2021. https://doi.org/10.3843/GLOWM.413113.
[52] Burlinson CEG, Sirounis D, Walley KR, Chau A. Sepsis in pregnancy and the puerperium. Int J Obstet Anesth 2018;36:96–107. https://doi.org/10.1016/j.ijoa.2018.04.010.
[53] Fan SR, Liu P, Yan SM, Huang L, Liu XP, et al. New concept and management for sepsis in pregnancy and the puerperium. Maternal–Fetal Med 2020;2(4):231–9. https://doi.org/10.1097/FM9.0000000000000058.
[54] Blauvelt CA, Nguyen KC, Cassidy AG, Gaw SL. Perinatal outcomes among patients with sepsis during pregnancy. JAMA Netw Open 2021;4(9):e2124109. https://doi.org/10.1001/jamanetworkopen.2021.24109.
[55] Royal College of Obstetricians and Gynaecologists. Bacterial sepsis in pregnancy. Green-Top guideline No. 64a. London: RCOG; 2012. https://www.rcog.org.uk/en/guidelines-research-services/guidelines/gtg64a/.
[56] Bright RA, Lima FV, Avila C, Butler J, Stergiopoulos K. Maternal heart failure. J Am Heart Assoc 2021;10(14):e021019. https://doi.org/10.1161/JAHA.121.021019.
[57] Cauldwell M, Chen KK, Lowe S, et al. Pregnancy in the woman with cardiovascular disease – congenital and acquired. In: Chen KK, editor. The continuous textbook of women's medicine series – obstetrics module, 8. Global Library of Women's Medicine; 2021. https://doi.org/10.3843/GLOWM.415373.
[58] Mendelson MA. Acquired heart disease in pregnancy. Glob. Libr. Women's Med; 2008. p. 1756–2228. https://doi.org/10.3843/GLOWM.10160.
[59] Schaufelberger M. Cardiomyopathy and pregnancy. Heart 2019;105:1543–51. https://doi.org/10.1136/heartjnl-2018-313476.
[60] Anthony J, Sliwa K. Decompensated heart failure in pregnancy. Card Fail Rev 2015;1(2):20–6. https://doi.org/10.15420/cfr.2015:24:2.

[61] Iftikhar SF, Biswas M. Cardiac disease in pregnancy. In: StatPearls [Internet]. Treasure Island (FL): StatPearls Publishing; 2023.

[62] Rojas-Suarez J, Bello-Muñoz C, et al. The continuous textbook of women's medicine series – obstetrics module. In: María Fernanda EV, editor. Obstetric emergencies. 13. Global Library of Women's Medicine; 2021. https://doi.org/10.3843/GLOWM.413803.

[63] Orfanoudaki IM, Obstetrics Gynecology, University of Crete, Greece. Review article: Pulmonary embolism in pregnancy: suspicion, diagnosis and therapy. Obstet Gynecol Int J 2019;10(1):1–13.

[64] Simcox LE, Ormesher L, Tower C, et al. Pulmonary thrombo-embolism in pregnancy: diagnosis and management. Breathe 2015;11:282–9.

[65] Dado CD, Levinson AT, Bourjeily G. Pregnancy and pulmonary embolism. Clin Chest Med 2018; 39(3):525–37. https://doi.org/10.1016/j.ccm.2018.04.007.

[66] Hobohm L, Farmakis IT, Münzel T, Konstantinides S, Keller K. Pulmonary embolism and pregnancy—challenges in diagnostic and therapeutic decisions in high-risk patients. Front. Cardiovasc. Med. 2022;9:856594. https://doi.org/10.3389/fcvm.2022.856594.

[67] Greer I, Thomson AJ. Green-top guideline No. 37b – thromboembolic disease in pregnancy and the puerperium: acute management. London: Royal College of Obstetricians and Gynaecologists; 2015.

[68] Becattini C, Agnelli G, Vedovati MC, et al. Multidetector computed tomography for acute pulmonary embolism: diagnosis and risk stratification in a single test. Eur Heart J 2011;32(13):1657–63.

[69] Winer -Muram HT, Boone JM, Brown HL, et al. Pulmonary embolism in pregnant patients:fetal radiation dose with helical CT. Radiology 2002;224:487–92.

CHAPTER 6

Acute kidney injury in a pregnant woman

Payel Bose[1], Ahsina Jahan Lopa[2] and Saurabh Debnath[3]
[1]Critical Care Medicine, Medica Superspecialty Hospital, Kolkata, West Bengal, India; [2]Department of ICU and Emergency, Brahmanbaria ICU Specialized Hospital, Dhaka, Bangladesh; [3]Critical Care Medicine, B P Poddar Hospital and Research Ltd., Kolkata, West Bengal, India

Introduction

Pregnancy-related acute kidney injury (Pr-AKI) is a heterogeneous disease entity that occurs due to a multitude of underlying etiologies. Pr-AKI is a complex challenge for the treating physician. The renal system undergoes anatomical and physiological changes during a normal pregnancy. Clinicians need to be familiar with these changes because they may affect the diagnosis and management of AKI. Renal plasma flow increases by 50%–70% during pregnancy in the first two trimesters [1]. Additionally, the glomerular filtration rate (GFR) increases from an average of 97 mL/minutes to 128 mL/mm by the end of the first trimester. Consequently, both blood urea nitrogen (BUN) and creatinine concentrations are lower than the normal range, and thus a normal BUN or creatinine level in a pregnant female may actually indicate underlying renal disease [2].

Epidemiology of Pr-AKI and maternal/fetal/renal outcomes

Pr-AKI has always been an important public health issue in developing countries. Recent reports from India have revealed a sharp decline in the proportion of Pr-AKI among hospitalized patients with acute kidney injury (AKI), from 15% in the 1980s to 1.5% in the 2010s; however, 30% of the recent cases were severe and required dialysis [3]. Most cases of Pr-AKI now occur in the postpartum rather than the postabortal period, reflecting a decline in septic abortions and the need for further improvement in peripartum care. Recent studies from China and India report maternal mortality rate associated with Pr-AKI of 4.0% and 5.8%, respectively, as compared with a rate of 20% during the 1980s [4,5]. These reports of the decrease incidence in maternal mortality associated with Pr-AKI in the developing countries are very encouraging, but are still unacceptably high in absolute numbers. Aggressive public health initiatives to deliver high-quality obstetric care to the most vulnerable sections of the population are needed to mirror the success achieved in the developed countries.

In the developed world, a significant decrease in Pr-AKI was noted by the end of the 20th century. An Italian study reported a decrease in Pr-AKI from one in 3000

The Kidney of the Critically Ill Pregnant Woman
ISBN 978-0-443-21473-8
https://doi.org/10.1016/B978-0-443-21473-8.00015-X

pregnancies in the 1960s to one in 18,000 pregnancies in the 1990s [6]. However, the landscape of Pr-AKI in the developed world is changing, with recent studies from Canada and the United States reporting an increased incidence of Pr-AKI. Much of the increase in diagnosis has been attributed to the different coding of AKI (ascertainment bias) and increased surveillance during the study period rather than an absolute increase [7]. This view is supported by data that most Pr-AKI cases reported over a 5-year period were minor and transient (87%) [8]. Although ascertainment bias may certainly be a contributing factor, there is also cause for real concern, as evidenced by the increase in severe Pr-AKI requiring dialysis. Furthermore, there were increased rates of severe maternal morbidity defined by Pr-AKI, shock, acute myocardial infarction, and respiratory distress syndrome between 2008 and 2009 as compared with 1998 and 1999 [9]. The reasons for these hard outcomes were attributed to factors such as older maternal age, pregnancies with hypertensive disorders, and underlying chronic kidney disease [10,11]. These reports are indicative of the new challenges facing the developed world, and highlight the need for ongoing studies to parse the differences in the increase of Pr-AKI due to ascertainment bias from the subset of Pr-AKI that are associated with increased maternal risk factors and maternal mortality.

Pr-AKI is also associated with significant fetal mortality and morbidity. The odds of perinatal mortality increase 3.4-fold when compared with pregnancies without Pr-AKI [12]. Studies from India have reported high perinatal mortality of 20%—45% due to intrauterine death, stillbirth, and prematurity. Long-term renal outcomes have not been well studied in women with Pr-AKI. In the short term, less severe Pr-AKI demonstrates favorable renal recovery at 40%—75% [13]. In contrast, 4%—9% of women with severe Pr-AKI remained dialysis dependent at 4—6 months postpartum [14]. The rate of progression to end-stage renal disease from Pr-AKI, in general, ranges from 1.5% to 2.5% [12].

Diagnosis of Pr-AKI

The definition of Pr-AKI used in literature is variable, ranging from an increase in serum creatinine to AKI needing dialysis [15]. Hemodynamic and vascular changes in normal pregnancy result in a 40%—50% increase in GFR. Thus, serum creatinine that is within the normal range for the general population could reflect significant compromise in renal function in a pregnant woman. In the general population, the RIFLE (Risk, Injury, Failure, Loss, and End Stage) and AKIN (Acute Kidney Injury Network) criteria are commonly used to define and classify AKI but are not well validated in pregnancy. Nevertheless, recent studies using the RIFLE and AKIN criteria report that most cases of Pr-AKI are of the AKIN stage 1 category and that a higher RIFLE category was associated with worse outcomes [16]. Although more studies to validate these criteria in Pr-AKI are needed, they can provide much needed uniformity.

Preeclampsia/HELLP

Preeclampsia is defined as blood pressure >140/90 mm Hg with proteinuria of more than 300 mg/days after 20 weeks of gestation in a previously normotensive woman or without proteinuria when there is evidence of end-organ damage [17]. It occurs in 2%–8% of pregnancies, most commonly in the second and third trimester period, but it can also occur in the postpartum period in up to 5% of cases [18,19]. Interestingly, a history of recovered AKI unrelated to pregnancy is associated with a higher risk of preeclampsia in future pregnancies [20]. Although preeclampsia is associated with a 30%–40% reduction in renal blood flow and GFR compared with a normal pregnancy, Pr-AKI is an uncommon manifestation (1%), unless preeclampsia is severe or associated with the HELLP syndrome [21]. The HELLP syndrome, associated with Pr-AKI in 7%–36% of cases, is classified as a preeclampsia/eclampsia continuum, even though 20% of cases do not have antecedent hypertension or proteinuria [22]. In practice, the differentiation of preeclampsia/HELLP from other causes, such as TTP/aHUS and lupus nephritis, is challenging due to overlapping clinical and laboratory parameters. Defects in placentation and maternal susceptibility play a central role in the development of preeclampsia, with placental derived angiogenic factors, such as soluble fms-like tyrosine kinase-1 (sFlt-1), placental growth factor, and soluble endoglin as key mediators of the disease [23]. These factors are being explored as biomarkers for preeclampsia diagnosis and its differentiation from other causes of Pr-AKI, as well as targets for therapeutic intervention [24]. An open clinical trial of sFlt-1 removal by dextran sulfate apheresis in 11 women with preeclampsia showed a reduction in proteinuria and prolongation of pregnancy by 2–21 days [25].

Complement system dysregulation has emerged as another possible underlying cause of maternal susceptibly and a mechanistic pathway for organ injury in preeclampsia/HELLP [26]. Mutations in complement regulatory genes have been reported in 8.5%–18.0% of women who developed preeclampsia in the setting of systemic lupus erythematosus (SLE) and/or antiphospholipid antibodies [27]. Single nucleotide polymorphisms in complement gene C3 can modify susceptibility to severe preeclampsia with both predisposing and protective effects based on the allele combination [28]. An in vitro study demonstrated that serum from women with preeclampsia/HELLP is cytotoxic due to complement activation and that administration of eculizumab, a complement inhibitor, ameliorates this finding [29].

Thrombotic microangiopathies: TTP/aHUS

The renal thrombotic microangiopathies comprise disorders such as TTP and aHUS. Their common characteristics are that of disseminated occlusion of arterioles and capillaries with fibrin and possibly agglutinated platelets resulting in mechanical destruction of red blood cells and decreased platelet count. Deficiency of ADAMTS-13, a von

Willebrand factor—cleaving protease, is responsible for most cases of TTP, occurring mostly in the second and third trimesters of pregnancies [30—32]. Pregnancy, itself a pro-coagulant state, is a trigger for TTP, especially in the setting of ADAMTS-13 deficiency. Hence, treatment of TTP requires plasma exchanges or fresh frozen plasma infusions for clearance of autoantibodies and restoration of enzymatic activity [33]. aHUS, on the other hand, is due to the excessive activation of the alternative complement pathway. Genetic mutations in the complement regulatory proteins, such as complement factor H, complement factor I, C3, membrane cofactor protein, or a combination, can result in an unchecked activation of alternative complement pathway [34]. A large European study recently showed that of 87 women who developed pregnancy-associated HUS, 56% exhibited novel or rare variants in complement genes, the most common mutations being in genes for complement factor H and complement factor I [35]. Those with complement gene variants fared worse: they required dialysis at presentation more frequently than those without, progressed to end-stage renal disease more frequently, and had a higher risk of relapse. They also demonstrated worse pregnancy outcomes as compared with women without detectable genetic defects: fetal loss at 4.8% versus 0% and preeclampsia at 7.7% versus 0%, respectively [36].

Similar to TTP, empiric plasma exchange has been the cornerstone of aHUS treatment: removal of autoantibodies to alternative complement pathway and replacement of deficient gene products with plasma. This is a reasonable approach given the extreme difficulty in differentiating cases with new-onset aHUS from TTP initially and the long turnaround time for genetic testing. However, it is not a definitive treatment, and it has shown to be effective in about 50% of adults [36]. The advent of eculizumab, a humanized monoclonal anti-C5 antibody that binds with high affinity to C5, thereby inhibiting the C5 cleavage and generation of the membrane attack complex, has revolutionized the treatment of aHUS [37]. Its use appears to be safe in pregnancy, as eculizumab has not been detected in umbilical cord or blood samples of neonates [38]. In an observational study of 22 women with pregnancy-associated HUS in which 10 patients or 45% of patients were treated with eculizumab, all treated patients showed a favorable clinical response [39]. Another unanswered question is whether eculizumab should be prophylactically administered in subsequent pregnancies in women with prior episodes of pregnancy-associated HUS. Again, there is no specific recommendation to do so, as the pregnancy history is not predictive of recurrence. However, it is recommended to follow these women very closely throughout their pregnancies until up to 3 months postpartum and to administer eculizumab immediately in case of recurrence.

Acute fatty liver of pregnancy

AFLP is a rare (1:20,000 pregnancies) but potentially fatal obstetric emergency, with maternal and neonatal mortality of 2% and 10%, respectively [40]. Mutations in maternal and fetal long-chain 3-hydroxyacyl-CoA dehydrogenase enzyme result in accumulation of long-chain fatty acids, which causes hepatotoxicity in the mother [41]. When it is severe, liver dysfunction results in elevated bilirubin, transaminitis, coagulopathy, hypoglycemia, lactic acidosis, ascites, and encephalopathy [42]. Liver biopsy, often impractical due to coagulopathy and not mandatory for diagnosis, reveals a hallmark histology finding of microvesicular fatty infiltration of hepatocytes [43]. Pr-AKI associated with AFLP is common, occurring with varying degrees of severity in 50%–75% of patients with AFLP [44]. Similar to liver pathology findings, an autopsy series reported free fatty acid deposition in the renal tubular epithelium in some cases [45]. Distinguishing AFLP from preeclampsia/HELLP is challenging, and in fact 20%–40% patients with AFLP have a concomitant diagnosis of preeclampsia/HELLP [46].

Practice guidelines from the American College of Gastroenterology recommend prompt delivery in women with AFLP, molecular testing for long-chain 3-hydroxyacyl- CoA dehydrogenase in mothers and offspring, and monitoring the offspring for clinical manifestation of long-chain 3-hydroxyacyl-CoA dehydrogenase deficiency, such as hypoketotic hypoglycemia and fatty liver [47]. Although most cases have spontaneous liver recovery after delivery, in severe cases, plasma exchange and liver transplantation have been successfully used [48]. Short-term renal outcome is also favorable, with normalization of laboratory findings after delivery, but long-term effects of AFLP are not well understood [49].

Lupus nephritis

Because SLE affects mostly women of childbearing age, the risk of developing lupus nephritis and subsequent Pr-AKI is a concern. Complications of lupus during pregnancy include lupus flares, preeclampsia, thrombotic events, and/or fetal complications, such as miscarriage, preterm birth, intrauterine growth restriction, and neonatal lupus [50]. Women with SLE/lupus nephritis are advised to undergo preconception counseling in a multidisciplinary approach. It is difficult to predict the course of lupus during pregnancy, as it is unclear if pregnancy predisposes to lupus flares or worsening of lupus nephritis [51]; however, the risk of flares does correlate with preconception lupus activity [52], low C3, and a history of lupus nephritis [53].

It may be possible to predict adverse pregnancy outcomes in SLE with angiogenic markers in the future: the PROMISSE (Predictors of Pregnancy Outcome: Biomarkers in Antiphospholipid Antibody Syndrome and Systemic Lupus Erythematosus) study demonstrated that women with the highest levels of sFlt-1 or a high sFlt-1:placental growth factor ratio in the late first trimester/second trimester exhibited an increased risk of an adverse outcome [54]. Lupus patients with concomitant antiphospholipid syndrome, in addition, have an increased risk of maternal complications (preeclampsia/HELLP) and fetal outcomes (fetal loss, intrauterine growth restriction, preterm labor)

[55]. Patients with SLE should be given low-dose aspirin preconceptionally or no later than gestational week 16 to prevent the development of preeclampsia and fetal growth restriction [56]. In patients with antiphospholipid syndrome, the addition of enoxaparin to low-dose aspirin is recommended to decrease rates of recurrent pregnancy loss [57]. If lupus nephritis is diagnosed during pregnancy, the mainstay of treatment includes steroids, calcineurin inhibitors, or azathioprine; on the other hand, mycophenolate mofetil and cyclophosphamide are avoided due to their known teratogenicity [58]. Hydroxychloroquine should be maintained throughout the pregnancy, as discontinuation has been associated with lupus flares [59].

Pyelonephritis

The occurrence of urinary tract infections is common in pregnancy due to functional, hormonal, and anatomical changes [60]. Untreated bacteriuria itself leads to low birth weight and preterm delivery, whereas its eradication reduces the incidence of pyelonephritis and improves fetal outcomes [61]. Thus, it is recommended by the American College of Obstetrics and Gynecology to conduct routine screening for bacteriuria at the first prenatal visit [62]. Risk factors of urinary tract infection include diabetes mellitus, HIV, urologic anomalies, sickle cell hemoglobinopathy, and history of *Chlamydia trachomatis* [63]. The most common culprit remains *Escherichia coli*, at an incidence of 76%–81%, whereas other gram-negative organisms, such as *Klebsiella*, *Enterobacter*, and *Proteus* species, occur at less than 10%; gram-positive organisms (group B *Streptococcus*) can occur at 10%. An untreated urinary tract infection can progress to pyelonephritis, which is a severe entity in pregnancy. Thus, early aggressive treatment with appropriate antibiotics is crucial and has significantly decreased the incidence of Pr-AKI from 20% to 2%. Hydration, on the other hand, must be carefully administered, as the risk of developing acute respiratory distress syndrome is high given the altered alveolar capillary membrane permeability due to endotoxin-mediated endothelial injury.

Antibiotic therapy should be directed by the microbiology and sensitivity analysis. Oral antibiotics, such as nitrofurantoin and beta-lactams, are preferred for uncomplicated urinary tract infections or cystitis. However, nitrofurantoin should be avoided in women with glucose-6-phosphate dehydrogenase deficiency and in the third trimester of all pregnancies due to the theoretical risk of hemolytic anemia in the fetus or newborn [64]. Another class of drugs to be cautious about is trimethoprim/sulfonamides: trimethoprim harbors concern of neural tube and cardiovascular defects in the first trimester, whereas sulfonamides carry the risk of increasing unbound bilirubin due to competitive binding causing fetal jaundice [65].

Treatment of pyelonephritis, on the other hand, requires the use of parenteral antibiotics. Avoidance of fluoroquinolones and aminoglycosides, if possible, is recommended owing to concerns over animal reports of arthropathy of the fetus in the former class [66] and the risk of nephrotoxicity of the mother in the latter.

Postrenal Pr-AKI

Obstructive causes of Pr-AKI are unusual in pregnancy. Physiologic hydronephrosis, which occurs in up to 90% of pregnant women, is a normal occurrence in pregnancy due to the compression of the ureter at the pelvic brim by the growing uterus and smooth muscle relaxation induced by increased progesterone levels [67]. The combination of urinary stasis and lithogenic factors of pregnancy (increased urinary calcium, oxalate, uric acid, and sodium) may contribute to renal calculi formation [68]. Bilateral renal calculi, although a rare cause of Pr-AKI, will almost always require a surgical intervention, such as ureteroscopy [69], placement of ureteral stents, or nephrostomy tubes [70]. Iatrogenic injuries to the bladder and ureters are exceptionally rare causes of Pr-AKI and are usually a result of emergent cesarean deliveries, especially in women with altered urologic anatomy, such as ectopic kidneys or duplication of ureters [71]. Obstructive uropathy, by uterine compression of the ureters, is another rare event that has shown to result in a high fetal mortality of 33% [70]. The etiologies ranged from multiple gestations, solitary kidney, polyhydramnios, and nephrolithiasis, and treatment was directed at the underlying cause [70].

Management of Pr-AKI

Successful management of Pr-AKI requires a multidisciplinary approach with close collaboration among nephrologists, obstetricians, intensivists, and other team members. The identification of the underlying etiology of Pr-AKI is crucial in its proper management.

The optimal care of pregnant women with AKI requires a multidisciplinary team, including a maternal–fetal medicine specialist, who should be involved early in the care. Early treatment for AKI typically includes optimization of the hemodynamic status and correction of any volume deficit. This will have a salutary effect on kidney function, helping to minimize progression of further kidney injury, and potentially facilitating recovery from AKI with a minimization of any residual chronic impairment of kidney function [72]. The intravascular volume and mean arterial pressure should be maintained, and acidosis and electrolyte abnormalities should be corrected. Further damage should be prevented by avoiding nephrotoxic agents. All medications need to be adjusted for the level of renal function. Pregnant women with AKI need to be monitored for any other organ dysfunction.

For biopsy-proven glomerulonephritis, steroid and immunosuppressive therapy will be necessary.

For cases of severe preeclampsia/HELLP syndrome or AFLP, prompt delivery of the fetus is warranted.

The use of i.v. magnesium remains a cornerstone in the prevention of seizures in a patient with preeclampsia, but as magnesium is renally excreted, women with moderate to severe Pr-AKI are at risk of magnesium toxicity. Monitoring of serum magnesium level (therapeutic levels of 5–8 mg/dL) and clinical toxicity (hypotension, hyporeflexia, and somnolence) is recommended during infusion.

For most cases of TTP, plasma exchange is recommended, whereas a diagnosis of aHUS will require the addition of eculizumab. Additional measures, such as volume resuscitation, prevention of further injury, timely initiation of renal replacement therapy, and prompt delivery of fetus may be needed. Resuscitation efforts should be carefully monitored, as women with either endotoxin-mediated injury or preeclampsia are prone to developing pulmonary edema. Most complications of Pr-AKI can be medically treated similar to nonpregnant patients: volume overload can be treated with loop diuretics, hyperkalemia managed with cation exchange resins, metabolic acidosis with alkali therapy, and anemia with blood transfusion and/or erythropoietin-stimulating agents. If despite these interventions, the kidney injury progresses or the patient develops signs of uremia, initiation of renal replacement therapy will be necessary. There are limited data on the timing of initiation, duration of therapy, or choice of modality of renal replacement therapy in Pr-AKI, thus the dialysis prescription needs to be individualized. Clinicians should be prepared to start dialysis early in pregnant patients and maintain effective doses as the pregnancy continues [73]. Acute indications for dialysis are the same for pregnant patients as for nonpregnant patients in terms of fluids, electrolytes, and complications of uremia; however, the replacement therapy should be initiated early and not postponed until the onset of advanced uremic manifestations [73,74]. Information from registries suggests that aggressive control of azotemia results in better fetal and maternal outcomes; therefore, we should target a BUN within the normal range. The options for renal replacement therapy in patients with AKI include peritoneal dialysis (PD), intermittent hemodialysis, and continuous renal replacement therapy (CRRT). PD is less efficient than hemodialysis in providing adequate clearance in an acute state, and it is more difficult to insert a PD catheter [75]. The gravid uterus will also limit the volume of fluid used in each exchange [76]. Despite these limitations, this treatment may be the only option available in many rural areas in developing countries. Adequate control of azotemia with a lower blood flow rate to avoid hemodynamic instability can be performed by either slower solute removal (over 6–12 hours by sustained low-efficiency dialysis) or by CRRT. Before acute hemodialysis is initiated, vascular access must be established. Placement of the venous dialysis catheter must be considered carefully, particularly in critically ill pregnant women. The location depends upon whether the patient has a clinical and ambulatory status [77]. The internal jugular vein is most often the preferred site. The femoral vein is another site that is used but is less preferred because of the increased risk of deep vein thrombosis during pregnancy.

Determining the ultrafiltration [78] goals in pregnant women can be challenging. The intravascular volume in a normal pregnancy is expanded; therefore, a guided UF that avoids intradialytic hypotension is critical. Monitoring uteroplacental perfusion, fetal blood flow, and amniotic fluid volume may assist in maintaining adequate fetal perfusion and titrating ultrafiltration [79,80]. This ultrafiltration can be achieved by minimizing the UF rate requirements by increasing the frequency and duration of treatments, as well as by ultrafiltration profiling. Vasoconstriction due to a lower body temperature has been used to increase vascular resistance and improve hemodynamic stability during dialysis. Hypothermia, however, may be an undesirable consequence in critically ill pregnant patients due to the adverse effects of vasoconstriction and possible decline in uterine perfusion [81]. Similarly, sodium profiling should be avoided because the serum sodium concentration in pregnant women is physiologically decreased, and the impact of sodium profiling on a fetus is not known. The dialyzate calcium and potassium concentration should be adjusted to avoid electrolyte disturbances.

The coagulation factors are also altered during pregnancy. Fibrinogen (factor I) is markedly increased by approximately 2 g/L at term [82]. Factors VII, VIII, and X are also increased. Furthermore, the activity of factors IX and XII increase, and factors XI and XIII decrease by 30% or more at term [83]. Antithrombin II probably decreases [84]. A progressive increase occurs in fibrin and fibrinogen degradation products throughout pregnancy. The agent of choice is heparin because it does not cross the placental barrier, has no teratogenicity, and is easy to dose [85]. Heparinization of the extracorporeal circuit (regional heparinization) with the use of its protamine sulfate is another method that is recommended to correct the excess anticoagulant effect of heparin, particularly in women with active bleeding [86]. Trisodium citrate safety in pregnancy is not known; therefore, its use should be avoided. The dialyzate bicarbonate concentration should vary based upon the acid–base status of the patient; however, overcorrection should be avoided, and the physiologic metabolic acidosis should be targeted.

Along with management of the mother, fetal monitoring for well-being and appropriate growth is an integral part of the care plan. If the gestation is less than 34 weeks and delivery is imminent, administration of glucocorticoids is indicated to reduce the risk of neonatal respiratory distress syndrome [87].

Fetal monitoring is needed for all pregnant women with AKI. In cases where the gestational age is 25 weeks or less, assessment is usually limited to Doppler auscultation to document the fetal heart [88]. At 26 weeks or more, other more sensitive methods can be used, including a fetal movement assessment, contraction stress test, nonstress test, biophysical profile, or umbilical artery Doppler velocimetry. Regular fetal monitoring with contingency planning for a possible emergency cesarean delivery is conducted if fetal distress is observed for fetuses that have reached an adequate level of maturity.

Conclusion

In conclusion, there remains a need to increase the awareness of pregnancy-related AKI. Pr-AKI is not a rare entity. Pr-AKI significantly increases the risk of fetal or maternal complications, both during the entire pregnancy and postpartum. A better understanding of the range of different causes driving renal function deterioration during pregnancy and prompt appropriate treatment of Pr-AKI could significantly improve the prognosis for mother and child, but this requires the involvement of an interdisciplinary team. Regular monitoring of kidney function should be done by kidney health providers during pregnancy, especially in women who are at higher risk. Pregnancy-related AKI continues to remain underrecognized and therefore needs continued exploration to ensure that long-lasting effects, including development of chronic kidney disease and need for kidney replacement therapy, are minimized.

References

[1] Koetje PM, Spaan JJ, Kooman JP, Spaanderman ME, Peeters ll. Pregnancy reduces the accuracy of the estimated glomerular filtration rate based on Cockroft-Gault and MDRD formulas. Reprod Sci 2011; 18(5):456–62.
[2] Dunlop W, Davison JM. Renal haemodynamics and tubular function in human pregnancy. Baillieres Clin Obstet Gynaecol 1987;1(4):769–87.
[3] Mahesh E, Puri S, Varma V, et al. Pregnancy-related acute kidney injury: an analysis of 165 cases. Indian J Nephrol 2017;27:113–7.
[4] Prakash J, Pant P, Prakash S, et al. Changing picture of acute kidney injury in pregnancy: study of 259 cases over a period of 33 years. Indian J Nephrol 2016;26:262–7.
[5] Huang C, Chen S. Acute kidney injury during pregnancy and puerperium: a retrospective study in a single center. BMC Nephrol 2017;18:146.
[6] Stratta P, Besso L, Canavese C, et al. Is pregnancy-related acute renal failure a disappearing clinical entity? Ren Fail 1996;18:575–84.
[7] Lunn MR, Obedin-Maliver J, Hsu CY. Increasing incidence of acute kidney injury: also a problem in pregnancy? Am J Kidney Dis 2015;65:650–4.
[8] Gurrieri C, Garovic VD, Gullo A, et al. Kidney injury during pregnancy: associated comorbid conditions and outcomes. Arch Gynecol Obstet 2012;286:567–73.
[9] Hildebrand AM, Liu K, Shariff SZ, et al. Characteristics and outcomes of AKI treated with dialysis during pregnancy and the postpartum period. J Am Soc Nephrol 2015;26:3085–91.
[10] Mehrabadi A, Liu S, Bartholomew S, et al. Hypertensive disorders of pregnancy and the recent increase in obstetric acute renal failure in Canada: population based retrospective cohort study. BMJ 2014;349:g4731.
[11] Mehrabadi A, Dahhou M, Joseph KS, et al. Investigation of a rise in obstetric acute renal failure in the United States, 1999–2011. Obstet Gynecol 2016;127:899–906.
[12] Liu Y, Ma X, Zheng J, Liu X, et al. Pregnancy outcomes in patients with acute kidney injury during pregnancy: a systematic review and meta-analysis. BMC Pregnancy Childbirth 2017;17:235.
[13] Arrayhani M, El Youbi R, Sqalli T. Pregnancy-related acute kidney injury: experience of the nephrology unit at the University Hospital of Fez, Morocco. ISRN Nephrol 2013:109034. 2013.
[14] Krishna A, Singh R, Prasad N, et al. Maternal, fetal and renal outcomes of pregnancy-associated acute kidney injury requiring dialysis. Indian J Nephrol 2015;25:77–81.
[15] Jim B, Garovic VD. Acute kidney injury in pregnancy. Semin Nephrol 2017;37:378–85.
[16] Kamal EM, Behery MM, Sayed GA, Abdulatif HK. RIFLE classification and mortality in obstetric patients admitted to the intensive care unit with acute kidney injury: a 3-year prospective study. Reprod Sci 2014;21:1281–7.
[17] American College of Obstetricians and Gynecologists, Task Force on Hypertension in Pregnancy. Hypertension in pregnancy. Report of the American College of Obstetricians and Gynecologists' task force on hypertension in pregnancy. Obstet Gynecol 2013;122:1122–31.

[18] Duley L. The global impact of pre-eclampsia and eclampsia. Semin Perinatol 2009;33:130–7.
[19] Powles K, Gandhi S. Postpartum hypertension. CMAJ (Can Med Assoc J) 2017;189:E913.
[20] Tangren JS, Powe CE, Ankers E, et al. Pregnancy outcomes after clinical recovery from AKI. J Am Soc Nephrol 2017;28:1566–74.
[21] Kuklina EV, Ayala C, Callaghan WM. Hypertensive disorders and severe obstetric morbidity in the United States. Obstet Gynecol 2009;113:1299–306.
[22] Machado S, Figueiredo N, Borges A, et al. Acute kidney injury in pregnancy: a clinical challenge. J Nephrol 2012;25:19–30.
[23] Maynard SE, Min JY, Merchan J, et al. Excess placental soluble fms-like tyrosine kinase 1 (sFlt1) may contribute to endothelial dysfunction, hypertension, and proteinuria in preeclampsia. J Clin Invest 2003;111:649–58.
[24] Maynard SE, Thadhani R. Pregnancy and the kidney. J Am Soc Nephrol 2009;20:14–22.
[25] Thadhani R, Hagmann H, Schaarschmidt W, et al. Removal of soluble fms-like tyrosine kinase-1 by dextran sulfate apheresis in preeclampsia. J Am Soc Nephrol 2016;27:903–13.
[26] Lynch AM, Salmon JE. Dysregulated complement activation as a common pathway of injury in pre-eclampsia and other pregnancy complications. Placenta 2010;31:561–7.
[27] Salmon JE, Heuser C, Triebwasser M, et al. Mutations in complement regulatory proteins predispose to preeclampsia: a genetic analysis of the PROMISSE cohort. PLoS Med 2011;8:e1001013.
[28] Lokki AI, Kaartokallio T, Holmberg V, et al. Analysis of complement C3 gene reveals susceptibility to severe preeclampsia. Front Immunol 2017;8:589.
[29] Vaught AJ, Gavriilaki E, Hueppchen N, et al. Direct evidence of complement activation in HELLP syndrome: a link to atypical hemolytic uremic syndrome. Exp Hematol 2016;44:390–8.
[30] Cosmai EM, Puzis L, Tsai HM, Lian EC. Thrombocytopenic purpura and cardiomyopathy in pregnancy reversed by combined plasma exchange and infusion. Eur J Haematol 2002;68:239–42.
[31] Ducloy-Bouthors AS, Caron C, Subtil D, et al. Thrombotic thrombocytopenic purpura: medical and biological monitoring of six pregnancies. Eur J Obstet Gynecol Reprod Biol 2003;111:146–52.
[32] Fujimura Y, Matsumoto M, Kokame K, et al. Pregnancy induced thrombocytopenia and TTP, and the risk of fetal death, in Upshaw-Schulman syndrome: a series of 15 pregnancies in 9 genotyped patients. Br J Haematol 2009;144:742–54.
[33] Stella CL, Dacus J, Guzman E, et al. The diagnostic dilemma of thrombotic thrombocytopenic purpura/hemolytic uremic syndrome in the obstetric triage and emergency department: lessons from 4 tertiary hospitals. Am J Obstet Gynecol 2009;200:381.e1–6.
[34] Fakhouri F, Roumenina L, Provot F, et al. Pregnancy associated hemolytic uremic syndrome revisited in the era of complement gene mutations. J Am Soc Nephrol 2010;21:859–67.
[35] Bruel A, Kavanagh D, Noris M, et al. Hemolytic uremic syndrome in pregnancy and postpartum. Clin J Am Soc Nephrol 2017;12:1237–47.
[36] Noris M, Caprioli J, Bresin E, et al. Relative role of genetic complement abnormalities in sporadic and familial aHUS and their impact on clinical phenotype. Clin J Am Soc Nephrol 2010;5:1844–59.
[37] Fakhouri F, Hourmant M, Campistol JM, et al. Terminal complement inhibitor eculizumab in adult patients with atypical hemolytic uremic syndrome: a single-arm, openlabel trial. Am J Kidney Dis 2016;68:84–93.
[38] Servais A, Devillard N, Fremeaux-Bacchi V, et al. Atypical haemolytic uraemic syndrome and pregnancy: outcome with ongoing eculizumab. Nephrol Dial Transplant 2016;31:2122–30.
[39] Huerta A, Arjona E, Portoles J, et al. A retrospective study of pregnancy-associated atypical hemolytic uremic syndrome. Kidney Int 2018;93:450–9.
[40] Knight M, Nelson-Piercy C, Kurinczuk JJ, et al. UK Obstetric Surveillance System. A prospective national study of acute fatty liver of pregnancy in the UK. Gut 2008;57:951–6.
[41] Ibdah JA, Bennett MJ, Rinaldo P, et al. A fetal fatty-acid oxidation disorder as a cause of liver disease in pregnant women. N Engl J Med 1999;340:1723–31.
[42] Vigil-De Gracia P. Acute fatty liver and HELLP syndrome: two distinct pregnancy disorders. Int J Gynaecol Obstet 2001;73:215–20.
[43] Rolfes DB, Ishak KG. Acute fatty liver of pregnancy: a clinicopathologic study of 35 cases. Hepatology 1985;5:1149–58.

[44] Nelson DB, Yost NP, Cunningham FG. Acute fatty liver of pregnancy: clinical outcomes and expected duration of recovery. Am J Obstet Gynecol 2013;209:456.e1–7.

[45] Slater DN, Hague WM. Renal morphological changes in idiopathic acute fatty liver of pregnancy. Histopathology 1984;8:567–81.

[46] Sibai BM. Imitators of severe pre-eclampsia. Semin Perinatol. 2009;33:196–205.

[47] Tran TT, Ahn J, Reau NS. ACG clinical guideline: liver disease and pregnancy. Am J Gastroenterol 2016;111(2):176–94. quiz 196.

[48] Tang WX, Huang ZY, Chen ZJ, Cui TL, Zhang L, Fu P. Combined blood purification for treating acute fatty liver of pregnancy complicated by acute kidney injury: a case series. J Artif Organs 2012;15: 176–84.

[49] Liu J, Ghaziani TT, Wolf JL. Acute fatty liver disease of pregnancy: updates in pathogenesis, diagnosis, and management. Am J Gastroenterol 2017;112:838–46.

[50] Clowse ME, Magder LS, Witter F, Petri M. The impact of increased lupus activity on obstetric outcomes. Arthritis Rheum 2005;52:514–21.

[51] Petri M. The Hopkins Lupus Pregnancy Center: ten key issues in management. Rheum Dis Clin North Am. 2007;33:227–35.

[52] Kwok LW, Tam LS, Zhu T, et al. Predictors of maternal and fetal outcomes in pregnancies of patients with systemic lupus erythematosus. Lupus 2011;20:829–36.

[53] Buyon JP, Kim MY, Guerra MM, et al. Kidney outcomes and risk factors for nephritis (flare/de novo) in a multiethnic cohort of pregnant patients with lupus. Clin J Am Soc Nephrol 2017;12:940–6.

[54] Kim MY, Buyon JP, Guerra MM, et al. Angiogenic factor imbalance early in pregnancy predicts adverse outcomes in patients with lupus and antiphospholipid antibodies: results of the PROMISSE study. Am J Obstet Gynecol 2016;214:e108.e1–108.e14.

[55] Ruiz-Irastorza G, Crowther M, Branch W, Khamashta MA. Antiphospholipid syndrome. Lancet. 2010;376:1498–509.

[56] Roberge S, Nicolaides K, Demers S, et al. The role of aspirin dose on the prevention of preeclampsia and fetal growth restriction: systematic review and meta-analysis. Am J Obstet Gynecol 2017;216: 110–120.e6.

[57] Fischer-Betz R, Specker C. Pregnancy in systemic lupus erythematosus and antiphospholipid syndrome. Best Pract Res Clin Rheumatol 2017;31:397–414.

[58] Hladunewich MA, Bramham K, Jim B, Maynard S. Managing glomerular disease in pregnancy. Nephrol Dial Transplant 2017;32(Suppl. l_1):i48–56.

[59] Clowse ME, Magder L, Witter F, Petri M. Hydroxychloroquine in lupus pregnancy. Arthritis Rheum 2006;54:3640–7.

[60] Connolly A, Thorp Jr JM. Urinary tract infections in pregnancy. Urol Clin North Am 1999;26: 779–87.

[61] Romero R, Oyarzun E, Mazor M, et al. Meta-analysis of the relationship between asymptomatic bacteriuria and preterm delivery/low birth weight. Obstet Gynecol 1989;73:576–82.

[62] ACOG Educational Bulletin. Antimicrobial therapy for obstetric patients. Number 245, March 1998 (replaces no. 117, June 1988). American College of Obstetricians and Gynecologists. Int J Gynaecol Obstet 1998;61:299–308.

[63] Ovalle A, Levancini M. Urinary tract infections in pregnancy. Curr Opin Urol 2001;11:55–9.

[64] Gait JE. Hemolytic reactions to nitrofurantoin in patients with glucose-6-phosphate dehydrogenase deficiency: theory and practice. DICP 1990;24:1210–3.

[65] Repchinsky C, editor. Sulfamethoxazole/trimethoprim [product monograph]. Compendium of Pharmaceuticals and Specialties. The Canadian drug Reference for health Professionals. Ottawa, ON: Canadian Pharmacists Association; 2007. p. 2254–8.

[66] Ingham BB, Brentnall DW, Dale EA, et al. Arthropathy induced by antibacterial fused N-alkyl-4-pyridone-3-carboxylic acids. Toxicol Lett 1977;1:21–6.

[67] Marchant DJ. Effects of pregnancy and progestational agents on the urinary tract. Am J Obstet Gynecol 1972;112:487–501.

[68] Smith CL, Kristensen C, Davis M, Abraham PA. An evaluation of the physicochemical risk for renal stone disease during pregnancy. Clin Nephrol 2001;55:205–11.

[69] Semins MJ, Matlaga BR. Kidney stones during pregnancy. Nat Rev Urol 2014;11:163–8.
[70] Jena M, Mitch WE. Rapidly reversible acute renal failure from ureteral obstruction in pregnancy. Am J Kidney Dis 1996;28:457–60.
[71] Rajasekar D, Hall M. Urinary tract injuries during obstetric intervention. Br J Obstet Gynaecol 1997; 104:731–4.
[72] Karvellas CJ, Farhat MR, Sajjad I, Mogensen SS, leung AA, Wald R, et al. A comparison of early versus late initiation of renal replacement therapy in critically ill patients with acute kidney injury: a systematic review and meta-analysis. Crit Care 2011;15(1):R72.
[73] Okundaye I, Abrinko P, Hou S. Registry of pregnancy in dialysis patients. Am J Kidney Dis 1998; 31(5):766–73.
[74] Malik G, Al-harbi A, Al-Mohaya S, Dohaimi H, Kechrid M, Shetaia MS, et al. Pregnancy in patients on dialysis-experience at a referral center. J Assoc Physicians India 2005;53:937–41.
[75] Davenport A. Peritoneal dialysis in acute kidney injury. Perit Dial Int 2008;28(4):423–4. author reply 4-5.
[76] Briones-Garduno JC, Diaz de leon-Ponce MA, Rodriguez-Roldan M, Briones-Vega CG, Torres-Perez J. Peritoneal dialysis in obstetric patients. Cir Cir 2006;74(1):15–20.
[77] Oliver MJ, Edwards lJ, Treleaven DJ, Lambert K, Margetts PJ. Randomized study of temporary hemodialysis catheters. Int J Artif Organs 2002;25(1):40–4.
[78] Jensen D, Duffin J, Lam YM, Webb KA, Simpson JA, Davies GA, et al. Physiological mechanisms of hyperventilation during human pregnancy. Respir Physiol Neurobiol 2008;161(1):76–86.
[79] Flack NJ, Sepulveda W, Bower S, Fisk NM. Acute maternal hydration in third-trimester oligohydramnios: effects on amniotic fluid volume, uteroplacental perfusion, and fetal blood flow and urine output. Am J Obstet Gynecol 1995;173(4):1186–91.
[80] Bamberg C, Diekmann F, Haase M, Budde K, Hocher B, Halle H, et al. Pregnancy on intensified hemodialysis: fetal surveillance and perinatal outcome. Fetal Diagn Ther 2007;22(4):289–93.
[81] Jost CM, Agarwal R, Khair-el-Din T, Grayburn PA, Victor RG, Henrich W. Effects of cooler temperature dialysate on hemodynamic stability in "problem" dialysis patients. Kidney Int 1993;44(3): 606–12.
[82] Manten GT, Franx A, Sikkema JM, Hameeteman TM, Visser G, de Groot PG, et al. Fibrinogen and high molecular weight fibrinogen during and after normal pregnancy. Thromb Res 2004;114(1): 19–23.
[83] Bremme KA. Haemostatic changes in pregnancy. Best Pract Res Clin Haematol 2003;16(2):153–68.
[84] Thornton P, Douglas J. Coagulation in pregnancy. Best Pract Res Clin Obstet Gynaecol 2010;24(3): 339–52.
[85] James KV, Lohr JM, Deshmukh RM, Cranley JJ. Venous thrombotic complications of pregnancy. Cardiovasc Surg 1996;4(6):777–82.
[86] Ginsberg JS, Kowalchuk G, Hirsh J, Brill-Edwards P, Burrows R. Heparin therapy during pregnancy. Risks to the fetus and mother. Arch Intern Med 1989;149(10):2233–6.
[87] ACOG Committee Opinion No. 475: antenatal corticosteroid therapy for fetalmaturation. Obstet Gynecol 2011;117:422–4.
[88] ACoG Practice Bulletin. Clinical management guidelines for Obstetrician-Gynecologists, Number 70, December 2005 (replaces practice Bulletin Number 62, May 2005). Intrapartum fetal heart rate monitoring. Obstet Gynecol 2005;106(6):1453–60.

CHAPTER 7

Preeclampsia and HELLP syndrome

Nehad Nabeel Al Sherawi[1], Vinay Singhal[2] and Utpal Sarma[3]
[1]Al Fujairah Hospital, EHA, United Arab Emirates; [2]Critical Care & Emergency Services, SPS Hospitals, Ludhiana, Punjab, India; [3]Critical Care, Sri Balaji Action Medical Institute, Paschim Vihar, New Delhi, India

Introduction

Preeclampsia (PE), eclampsia, and HELLP syndrome are the hypertensive disorders of pregnancy that led major maternal and perinatal morbidity and mortality. Hypertension in pregnancy is defined as blood pressure more than 140/90 mm of Hg on two different occasions, 4 hours apart. It can present as chronic hypertension, gestational hypertension, preeclampsia, eclampsia, or HELLP syndrome (**H**emolysis, **E**levated **L**iver enzymes, **L**ow **P**latelet counts) (Box 7.1). There is no definitive therapy for these severe forms of hypertensive diseases; however, current approach is to prevent, diagnose early and timely intervention may improve maternal and fetal outcomes. This chapter provides an update on the current management strategies for the preeclampsia and HELLP syndrome.

Preeclampsia

Multisystemic disorder arising after 20 weeks of gestation in a previously normotensive woman characterized by (Table 7.1):

1. New onset of hypertension (i.e., systolic blood pressure (SBP) $\geq$ 140 mmHg and/or diastolic blood pressure (DBP) $\geq$ 90 mmHg) and proteinuria ($\geq$300 mg/24 hours). OR

BOX 7.1

Preexisting/chronic hypertension	Blood pressure $\geq$140/90 mm Hg, before the 20th week of pregnancy or (if diagnosed in pregnancy) persists 6 weeks after delivery
Gestational hypertension	Hypertension developing after 20th week of pregnancy, settles 6 weeks after delivery
Preeclampsia (new onset or superimposed on preexisting hypertension)	Worsening control of hypertension, associated with new-onset proteinuria (300 mg/24 hours) or a doubling in any proteinuria measured quantitatively, during the first trimester, after the 20th week of pregnancy
Eclampsia	Characterized by convulsions with hypertension in pregnancy, develops hypertension following the convulsion
HELLP syndrome	Severe form of preeclampsia

The Kidney of the Critically Ill Pregnant Woman
ISBN 978-0-443-21473-8
https://doi.org/10.1016/B978-0-443-21473-8.00017-3

Table 7.1 Definitions of preeclampsia.

Feature	American College of Obstetricians and Gynecologists (ACOG)-2013	International Society for the Study of hypertension in pregnancy (ISSHP) – 2014
Blood pressure	- SBP $\geq$140 mmHg and/or DBP $\geq$90 mmHg on 2 occasions at least 4 hours apart after 20 weeks of gestation in a previously normotensive woman. - SBP $\geq$150 mmHg and/or DBP $\geq$110 mmHg confirmed within shorter interval (minutes).	New onset of hypertension (>140 mmHg systolic or >90 mmHg diastolic) after 20 weeks gestation based on the average of at least two measurements, taken at least 15 minutes apart.
Proteinuria	- $\geq$300 mg in a 24-hour urinary collection. - Protein/creatinine ratio $\geq$0.3 (each measured in mg/dL) in a random urine sample. - Urine dipstick of 1+ (used only if other measures are not available). NB: Presence of proteinuria is not a must for diagnosis.	- Urine dipstick should be used as a screening test (1+ indicates significant proteinuria). - $\geq$ 0.3 g in a 24-hour urinary collection. - Protein/creatinine ratio $\geq$30 mg/mmol in a random urine sample. NB: Presence of proteinuria is not a must for diagnosis.
Other features	- In the absence of proteinuria, new onset hypertension associated with: **1.** Platelet count $\leq$100,000/mL. **2.** Elevated transaminase levels twice normal. **3.** Serum creatinine >1.1 mg/dL or doubling of serum creatinine in the absence of other renal disease. **4.** Pulmonary edema. **5.** Cerebral or visual problems.	- In the absence of proteinuria, new onset hypertension associated with: **1.** One or more adverse conditions or **2.** One or more of severe complications These may include: **(1)** Renal insufficiency (creatinine >90 μmol/L; 1.02 mg/dL) **(2)** Liver involvement (doubling of serum transaminases and/or severe right upper quadrant pain) **(3)** Neurological complications (eclampsia, altered mental status, blindness, stroke, or more commonly hyperreflexia when accompanied by clonus and severe headaches when accompanied by hyperreflexia and persistent visual scotomata) **(4)** (Hematological complications (platelet count <150,000/dL, DIC, and hemolysis)
Uteroplacental dysfunction (fetal growth restriction)	- Not required for diagnosis	- Required for diagnosis

2. New onset of hypertension and significant end-organ dysfunction with or without proteinuria. Organ dysfunction may be any of the following:
   - Liver function deterioration
   - Thrombocytopenia
   - Pulmonary edema
   - Renal impairment
   - Neurologic symptoms (visual sensations)

Preeclampsia with severe features (formerly severe preeclampsia)

Previously, severe preeclampsia was defined using the following criteria (1 of following criteria is required) (Table 7.2):

1. Blood pressure $\geq$160 mm Hg systolic or $\geq$110 mm Hg diastolic on two occasions at least 6 hours apart while patient is on bed rest
2. Proteinuria $\geq$5 g in 24-hour urine specimen or $\geq$3 g on two random urine samples collected at least 4 hours apart
3. Oliguria <500 mL in 24 hours
4. Cerebral or visual symptoms
5. Pulmonary edema or cyanosis
6. Epigastric or right upper quadrant pain
7. Impaired liver function
8. Thrombocytopenia
9. Fetal growth restriction

Currently, severe preeclampsia is renamed and consists of pregnant ladies who have severe hypertension (as defined above) and/or specific signs or symptoms of significant end-organ dysfunction that signify the severe end of the preeclampsia spectrum.

Resistant preeclampsia

Defined as the need for three antihypertensive medications for blood pressure control at $\geq$20 weeks of gestation.

Preeclampsia superimposed upon chronic hypertension

Defined as:

1. Preeclampsia occurs in a patient with preexisting chronic hypertension (primary or secondary hypertension that precedes pregnancy)
   OR
2. Hypertension is present on at least two occasions before the 20th week of gestation or persists longer than 12 weeks postpartum.

It is characterized by any of the following occurring after 20 weeks of gestation or postpartum in a patient with chronic hypertension:

Table 7.2 Adverse symptoms and severe complications of preeclampsia.

Organ system affected	Adverse symptoms	Severe complications
CNS	Headache/visual symptoms	Eclampsia PRES Retinal detachment Glasgow coma scale <13 Stroke, TIA
Cardiorespiratory	Chest pain Dyspnea Oxygen saturation (SpO_2) <97%	Uncontrolled severe hypertension (on >3 drugs) Fall in (SpO_2): Need >50% oxygen or intubation for respiratory reason, pulmonary edema Inotropic support Myocardial ischemia/infarction
Hematological	Elevated WBC count Raised INR or aPTT Low platelets	Platelets $<50 \times 10^9/L$ Transfusion of blood products
Renal	Elevated serum creatinine Elevated uric acid	Acute kidney injury New indication for hemodialysis
Hepatic	Nausea/vomiting Epigastric/right upper quadrant pain Elevated liver enzymes/bilirubin Low plasma albumin	Hepatic dysfunction (INR > 2) Hepatic hematoma/rupture
Fetoplacental	Nonreassuring fetal heart rate Intrauterine growth retardation Oligohydramnios	Maternal of fetal compromise (abruptio placenta) Stillbirth

- Worsening hypertension
- New onset of proteinuria or a sudden increase in proteinuria
- Significant new end-organ dysfunction

Eclampsia

Generalized tonic clonic seizures in the absence of other neurologic conditions that could account for the seizure, generally in addition to preeclampsia criteria.

Subtypes of preeclampsia.

There are several types of preeclampsia depending on onset of the disease, severity of hypertension, and other organ-system dysfunction and long-term prognosis. These subtypes represent different pathophysiological mechanisms leading to same clinical manifestations. Subtypes of preeclampsia include:

1. According to onset
   - Early onset (<34 weeks of gestation).
   - Intermediate onset (≥34 to <37 weeks of gestation).
   - Late onset (>37 weeks of gestation).

   Other resources classify preeclampsia into early onset (<34 weeks) and late onset (≥34 weeks) of gestation. Early onset preeclampsia is associated with more severe placental and maternal/fetal clinical findings and worse maternal/fetal outcomes. It is associated with an increase of growth-restricted infants and pathological evidence of placental malperfusion.

HELLP syndrome

Intravascular hemolysis, thrombocytopenia, and hematological abnormalities were first described in 1954 in patients with severe preeclampsia [1]. However, the term HELLP syndrome was coined by Weinstain in 1982 and he described this syndrome as complication of hypertension in pregnancy [2]. HELLP is seen in up to 10%–20% of women with severe preeclampsia, occurring in 0.2%–0.9% of all pregnancies [3]. About 70% of cases occur in the third trimester of pregnancy, and the remainder occurs within 48 hours of delivery [4]. Although HELLP has a different and distinct definition, it is often considered to be a severe form of PE. However, it challenged as preceding hypertension or proteinuria is absent in at least 15%–20% of patients with HELLP syndrome [5]. The clinical onset of HELLP may be rapid and associated with severe disease. Today obstetrician and intensivist have developed a clear understanding of this syndrome.

Diagnosis

Most patients with HELLP initially present with vague symptom like malaise, nausea, right upper quadrant, or epigastric pain and often misdiagnosed as viral prodrome. These symptoms are often accompanied by vomiting, headache, and blurred vision in varying combinations [6].

It is important to note that a delayed presentation of HELLP syndrome after delivery, despite absence of signs of preeclampsia at delivery, occurs in up to 30% of cases [4].

The diagnosis of HELLP syndrome is based on an assessment of the clinical features and on laboratory validation of clinical findings although there is no single diagnostic test that is specific for the disorder. Hemolysis in patients with HELLP is generally mild. Fragmented red blood cells (schistocytes) are seen on peripheral blood smears, and the serum LDH level is elevated. In addition, serum aminotransferase (AST) levels are also elevated minimally. However, sometime the level of AST becomes greater than 1000 U/L in association with laboratory signs of cholestasis. Serum bilirubin levels can often be mildly elevated, compatible with the finding of hemolysis. Although most pregnant women with low platelet counts and preeclampsia have HELLP syndrome, the differential diagnosis includes other causes of thrombocytopenia, which includes tropical fever in tropical area. Furthermore, immune thrombocytopenic purpura, thrombotic thrombocytopenic purpura (TTP), and the antiphospholipid antibody syndrome, which itself may be associated with early onset of HELLP are to be considered. Patient with TTP mainly present in the first trimester, unlike patients with HELLP syndrome. In addition, they have a characteristic pentad: fever, altered mental status, acute kidney injury/hematuria, hemolytic anemia, and thrombocytopenia. The severity of thrombocytopenia is more in TTP which may help to distinguish PE/HELLP from TTP. Approximately only 1% of patients with HELLP syndrome have platelets <30 × 10°L, while patients with acquired TTP have lower platelet counts, and while approximately 2% of patients with acquired TTP have platelet counts >30 × 10°L [7].

Pregnancy-related hemolytic uremic syndrome (HUS) presents similar to TTP clinically but without fever. Furthermore, ADAMTS 13 is not lowered unlike in TTP. Elevated serum aminotransferase levels in patients with preeclampsia are most frequently misdiagnosed as being caused by viral hepatitis.

Acute fatty liver of pregnancy (AFL) is appeared in third trimester, similar to patients with HELLP syndrome. These patients mainly have an obstructive pattern of liver injury unlike in patients with HELLP syndrome. HELLP syndrome should be differentiated from other disorders of pregnancy with similar features [8,9]. Patients with preeclampsia with DIC behave similar to HELLP syndrome. Thrombocytopenia and hemolytic anemia is present in systemic lupus erythematosus flare-up. However, the presence of typical characteristics of lupus-like malar rash, pleuritis, arthritis, and normal liver function test helps to differentiate it from a typical HELLP syndrome.

Antiphospholipid syndrome (APS) may cause confusion since patient with APS may have hemolysis and thrombocytopenia along with hypertension and proteinuria similar to HELLP syndrome. Clinical history will help to rule out dominant features of APS which are recurrent arterial/venous thrombosis and repeated spontaneous abortions <10 weeks gestation or fetal loss to differentiate from HELLP. The diagnosis of APS relies on persistent high titers of antiphospholipid antibodies (aPLs), mainly elevated lupus anticoagulant, anticardiolipin antibodies, and beta2-glycoprotein antibodies are needed for diagnosis [10]. A condition termed catastrophic APS develops in women with APLS and HELLP which results in an extensive microangiopathy and multiorgan failure [11].

Classification

In past, clinician relied on either on strict TENNESEE (Box 7.2) or simplified MISSISSIPI triple class classification (Box 7.3). More recently, the diagnostic criteria for HELLP

BOX 7.2

Tennessee classification

1. Microangiopathic hemolytic anemia (MAHA) with abnormal blood smear, low serum haptoglobin, and elevated serum LDH levels
2. Serum LDH level >600 IU/L or twice the laboratory upper limit of normal and serum AST level >70 IU/L or twice the laboratory upper limit of normal, or serum bilirubin level >1.2 mg/dL
3. Platelet count $<100{,}000/\mu L$

Incomplete HELLP syndrome is defined as the presence of only 1 or 2 of these abnormalities and may be less severe.

BOX 7.3

Mississippi triple-class classification.

Class I: platelet count nadir $\leq 50{,}000/mm^3$

Class II: platelet count nadir $>50{,}000/mm^3$ and $\leq 100{,}000/mm^3$

Class III: platelet count nadir $>100{,}000/mm^3$ and $\leq 150{,}000/mm$

BOX 7.4

1. Hemolysis and at least two of the following:
    a. Schistocytes and burr cells on peripheral smear
    b. Serum bilirubin ≥ 1.2 mg/dL
    c. Low serum haptoglobin
    d. Severe anemia unrelated to blood loss
2. Elevated liver enzyme levels:
    a. AST or ALT $\geq$ twice the upper limit of normal
    b. LDH $\geq$ twice the upper limit of normal
3. Platelets $<100{,}000/mm^3$

syndrome have been standardized by the American College of Obstetricians and Gynecologists Task Force in Hypertension in Pregnancy (Box 7.4).

Pathogenesis preeclampsia and HELLP

THE inflammatory response

Preeclampsia and HELLP syndrome are now thought to be consequences of abnormal placental formation. Consequently, there are failures of trophoblast invasion of the uterine lining and dilation of the spiral arteries which result the physiologic inability to increase uteroplacental perfusion appropriately as gestation progresses. It causes secondary release of placental products which is thought to initiate the clinical disease. The inflammatory response with coagulation cascade and complement activation is caused by syncytiotrophoblast particles (STBM) and other placental factors. These substances interact with maternal immune cells and vascular endothelial cells which causes activation of coagulation and complement factors [12,13]. As a result, there is high levels of activated leukocytes, inflammatory cytokines, TNF-α, and active Von Willebrand factor induces thrombotic microangiopathy with platelet–fibrin thrombi in microvessels [14].

Oxidative stress at placenta:

In response to poor placentation, there is generation of oxidative stress in the placenta which is further magnified in the maternal endothelium. It causes release of cytokine leading to cell injury, vasoconstriction, and subsequent activation of coagulation cascade.

Microangiopathic hemolytic anemia

Erythrocytes are damaged as they pass through blood vessels with impaired endothelium and fibrin strands, resulting in MAHA [11]. Therefore, schistocytes and/or burr cells are found on peripheral blood smear [15]. Lactate dehydrogenase (LDH) concentration may rise because of hemolysis. Free hemoglobin binds to unconjugated bilirubin in the spleen or haptoglobin in plasma [15].

The substances resulting from intravascular hemolysis promote the activation of coagulation cascade and increase DIC risk. A subset of HELLP syndrome is caused by complement dysregulation associated with thrombotic microangiopathy and can present with pregnancy-related HUS [16].

Liver and kidney dysfunctions

Microangiopathy seen in HELLP syndrome restricts portal blood flow which enhances hepatocyte damage. Hepatic hematoma and rupture complicating preeclampsia presumably result from extravasation of blood from one or several microscopic areas of periportal hemorrhage under Glisson capsule [17]. Kidney dysfunction is usually moderate in HELLP patients which might be due to thrombotic microangiopathy and acute tubular necrosis.

Complication of HELLP syndrome

The clinical abnormalities that characterize HELLP syndrome usually resolve rapidly after childbirth. Rarely, HELLP syndrome becomes gradually worse prior to delivery, with subsequent development of postpartum liver failure, sepsis, consumptive coagulopathy, and, rarely, even death. In the absence of appropriate supportive therapy and expedited delivery, affected patients may progress to renal failure, hepatic hematoma, and hepatic rupture. Postpartum HELLP is more likely to develop acute renal failure and pulmonary edema.

Laboratory findings to diagnose DIC are represented by platelet count, D-dimer, antithrombin, protein C, thrombin–antithrombin complex (TAT), and prothrombin time. The criteria to diagnose overt or uncompensated DIC require several abnormal findings in these parameters.

A moderately elevated D-dimer concentration is a nonspecific indicator of disease activity. TAT (thrombin antithrombin complex) assay value over 10 μg/L most specifically reflect the DIC process [18].

Predisposing factors and preventive measure of preeclampsia

Obesity, age more than 35 years, primiparity, previous history of preeclampsia, and West African population are at increased risk of developing preeclampsia. Multiparity women have increased morbidity and mortality.

It is important to screen the pregnant women during their antinatal visit and preventive measures are to be taken who are at high risk of developing preeclampsia. Low dose aspirin (75–100 mg) is effective in reducing the incidence of preeclampsia by 18% [19].

Ecosprin is to be started at the end of first trimester.

Calcium supplementation might reduce the incidence of preeclampsia especially when dietary intake of calcium is low [20].

Management

Most of the patients with preeclampsia gets benefited with the initiation of antihypertensive therapy. Primary goal of antihypertensive therapy is in reducing the risk of severe hypertension and other maternal complications. The target blood pressure should be 135/85 mm Hg or less. Commonly used antihypertensive agents to control blood pressure include labetalol, nifedipine, and methyldopa [21].

However, combined alpha and beta blocker (typically labetalol) and calcium-channel blockers (usually nifedipine) are more effective than other alternatives for preventing severe hypertension [22].

Intravenous agents (e.g., labetalol or hydralazine) can be used to control severe hypertension. However, it is important to monitor fetal heart-rate because of the risk of developing fetal bradycardia due to maternal hypotension.

In women with preeclampsia, magnesium sulfate reduces the risk of an eclamptic seizure by 58% [23].

Intravenous magnesium sulfate should be initiated when there is severe hypertension or has eclamptic fits. It important to keep the patient hydrated.

Patients with pregestation are advised to continue pregnancy till 37 weeks. They are closely monitored for any adverse features and if evident delivery of baby is expedited irrespective of gestation. For preeclampsia at 37 weeks' gestation or beyond, initiating birth is warranted.

Management of HELLP syndrome is primarily supportive; patients should be treated in an intensive care setting prior to delivery. Some affected patients may have a decline in serum aminotransferase levels and a rise in platelet counts with supportive care. Therefore, under such circumstances, a delay in delivery to achieve fetal maturity may be appropriate. Many affected women receive glucocorticoids before delivery, not as disease treatment but to promote fetal lung maturity; however, glucocorticoid therapy has also been used as therapy in this setting. Glucocorticoid therapy of HELLP syndrome may improve platelet counts and serum ALT levels and reduce hospital and ICU stays, but has not been associated with a significant improvement in maternal morbidity and mortality. However, in two large double-blind, randomized trials, there was no proven benefit to administering dexamethasone in patients with HELLP syndrome as it did not improve maternal or perinatal outcomes [24,25].

Patients with severe preeclampsia and HELLP syndrome may require antepartum platelet transfusions and hemodialysis. Plasmapheresis after delivery has been advocated for some patients to prevent postpartum thrombotic microangiopathy. Plasmapheresis is proposed as a therapeutic method in patients who present with a progressive increase in serum bilirubin and serum creatinine levels and have severe thrombocytopenia. It removes some pathological substances including ammonia, endotoxins, bilirubin, and inflammatory cytokines. Furthermore, plasmapheresis is also beneficial in decreasing the hemolysis marker, LDH, and increasing the level of platelets. Full recovery without sequelae is anticipated for the vast majority of affected patients. Prophylactic transfusion of platelets at delivery does not reduce the incidence of postpartum hemorrhage or hasten normalization of the platelet count.

Conclusion

Preeclampsia and HELLP syndrome are associated with major maternal and fetal morbidity and mortality. There is a need to improve antenatal care of expectants, so as to identify the disease early. There may be a challenge in differentiating HELLP syndrome from other pregnancy-induced disorders like fatty liver of pregnancy. There is no definitive therapy for HELLP syndrome; however, timely intervention including maternal stabilization and delivery may decrease the morbidity and mortality. Further studies are needed to address the role of corticosteroids and plasmapheresis in the management of HELLP syndrome.

References

[1] Pritchard JA, Weisman R, Ratnoff OD, et al. Intravascular hemolysis, thrombocytopenia and other hematologic abnormalities associated with severe toxemia of pregnancy. N Engl J Med 1954;250: 89–98.

[2] Weinstein L. Syndrome of hemolysis, elevated liver enzymes, and low platelet count: a severe consequence of hypertension in pregnancy. Am J Obstet Gynecol 1982;142(2):159–67. PMID 7055180.
[3] Rahman TM, Wendon J. Severe hepatic dysfunction in pregnancy. Q J Med 2002;95:343.
[4] Kirkpatrick CA. The HELLP syndrome. Acta Clin Belg 2010;65(2):91–7.
[5] Fitzpatrick KE, Hinshaw K, Kurinczuk JJ, et al. Risk factors, management, and outcomes of hemolysis, elevated liver enzymes, and low platelets syndrome and elevated liver enzymes, low platelets syndrome. Obstet Gynecol 2014;123(3):618–27.
[6] Sibai BM. Diagnosis, controversies, and management of the syndrome of hemolysis, elevated liver enzymes, and low platelet count. Obstet Gynecol 2004;103:981–99.
[7] Reese JA, Peck JD, Deschamps DR, et al. Platelet counts during pregnancy. N Engl Med 2018;379(1): 32–43. https://doi.org/10.1056/NEJMoa1802897.
[8] Wallace K, Harris S, Addison A, et al. HELLP syndrome: pathophysiology and current therapies. Curr Pharm Biotechnol 2018;19(10):816–26.
[9] Sibai BM. Imitators of severe preeclampsia. Obstet Gynecol 2007;109(4):956–66.
[10] Rao D, Chaudhari NK, Moore RM, et al. HELLP syndrome: a diagnostic conundrum with severe complications. BMJ Case Rep 2016;17(2016).
[11] Abildgaard U, Heimdal K. Pathogenesis of the syndrome of hemolysis, elevated liver enzymes, and low platelet count (HELLP): a review. Eur J Obstet Gynecol Reprod Biol 2013;166:117–23.
[12] May D, Djonov V, Zamir G, et al. A transgenic model for conditional induction and rescue of portal hypertension reveals a role of VEGF-mediated regulation of sinusoidal fenestrations. PLoS One 2011; 6:e21478.
[13] Braet F, Wisse E. Structural and functional aspects of liver sinusoidal endothelial cell fenestrae: a review. Comp Hepatol 2002;1:1.
[14] Endo D, Kogure K, Hasegawa Y, et al. Activin A augments vascular endothelial growth factor activity in promoting branching tubulogenesis in hepatic sinusoidal endothelial cells. J Hepatol 2004;40: 399–404.
[15] Sibai BM. Diagnosis, controversies, and management of the syndrome of hemolysis, elevated liver enzymes, and low platelet count. Obstet Gynecol 2004;103:981–91.
[16] Burwick RM, Feinberg BB. Eculizumab for the treatment of preeclampsia/HELLP syndrome. Placenta 2013;34(2):201–3.
[17] Tsokos M, Longauer F, Kardosová V. Maternal death in pregnancy from HELLP syndrome. A report of three medico-legal autopsy cases with special reference to distinctive histopathological alterations. Int J Leg Med 2002;116:50–3.
[18] Haram K, Svendsen E, Abildgaard U. The HELLP syndrome: clinical issues and management. A review. BMC Pregn Childb 2009;9:1–15.
[19] Duley L, Meher S, Hunter KE, et al. Antiplatelet agents for preventing pre-eclampsia and its complications. Cochrane Database Syst Rev 2019;2019:CD004659.
[20] Hofmeyr GJ, Lawrie TA, Atallah AN, et al. Calcium supplementation during pregnancy for preventing hypertensive disorders and related problems. Cochrane Database Syst Rev 2018;10:CD001059.
[21] National Institute for Health and Care Excellence. Hypertension in pregnancy: diagnosis and management. June 25, 2019. https://www.nice.org.uk/guidance/NG133. [Accessed 1 April 2020].
[22] Abalos E, Duley L, Steyn DW, et al. Antihypertensive drug therapy for mild to moderate hypertension during pregnancy. Cochrane Database Syst Rev 2018;10:CD002252.
[23] Altman D, Carroli G, Duley L, et al. Do women with pre-eclampsia, and their babies, benefit from magnesium sulphate? The Magpie Trial: a randomised placebo-controlled trial. Lancet 2002;359: 1877–90.
[24] Fonseca JE, Méndez F, Cataño C, et al. Dexamethasone treatment does not improve the outcome of women with HELLP syndrome: a double-blind, placebo-controlled, randomized clinical trial. Am J Obstet Gynecol 2005;193(5):1591–8.
[25] Katz L, de Amorim MM, Figueiroa JN, et al. Postpartum dexamethasone for women with hemolysis, elevated liver enzymes, and low platelets (HELLP) syndrome: a double-blind, placebo-controlled, randomized clinical trial. Am J Obstet Gynecol 2008;198(3):283.e1–8.

CHAPTER 8

Hemolytic uremic syndrome in pregnancy

Jomana Abdullah Al Mashikas[1] and Mohammed Alabdrab Alnabi[2]
[1]Department of Obstetrics & Gynecology, Qatif Central Hospital, Qatif, Saudi Arabia; [2]Department of Critical Care Medicine, Dammam Medical Complex, Dammam, Saudi Arabia

Definition and background

Hemolytic uremic syndrome (HUS) is a disorder that is defined simply as the triad of thrombocytopenia, microangiopathic hemolytic anemia, and acute kidney injury (AKI). Atypical HUS (aHUS) is a rare form of HUS, which accounts about 5%–10% of cases. Pregnancy is considered one of the risk factors that can lead to aHUS [3,6].

There is plenty of documentation in the literature confirming the link between some inherited genetic defects and the uncontrolled activation of the alternative complement pathway. This pathway is the basic pathophysiology of aHUS. It leads to endothelial damage and platelet activation, which blocks the distal vessels of the arterial system, leading to end-organ damage—see Fig. 8.1 [1,2]. Other possible triggers that may activate the complement pathway in genetically predisposed pregnant ladies are maternal–fetal bleeding, preeclampsia, systemic infection and inflammation, some drugs, and cancer [6].

There are more than 120 different mutations that account for more than half of the cases [3]. The most common involved genetic mutation (30%) is complement factor H (CFH); other involved mutations are membrane cofactor protein (MCP), complement factor I (CFI), C3, and complement factor B (CFB). It could also be a combined or a triple mutation [4].

Pregnancy-associated hemolytic uremic syndrome (Pa-HUS) accounts for 21% of adult female patients with aHUS [2]. One out of 25,000 pregnancies is affected by pa-HUS [6]. 86% of the cases occur in the postpartum period [5], and it has significant renal involvement. Despite a history of a normal first pregnancy in genetically predisposed women, they usually develop pa-HUS in their second pregnancy [2].

Diagnosis

The diagnostic criteria of aHUS are hemolytic anemia (anemia in the presence of broken red blood cells—schistocytes), low platelet count (thrombocytopenia), and kidney dysfunction, but genetic testing to identify the mutation linked to pa-HUS might not be available in many facilities, and, if available, will usually take a long time to perform (up to a few weeks). The genetic test is recommended to diagnose aHUS; any patient

The Kidney of the Critically Ill Pregnant Woman
ISBN 978-0-443-21473-8
https://doi.org/10.1016/B978-0-443-21473-8.00019-7

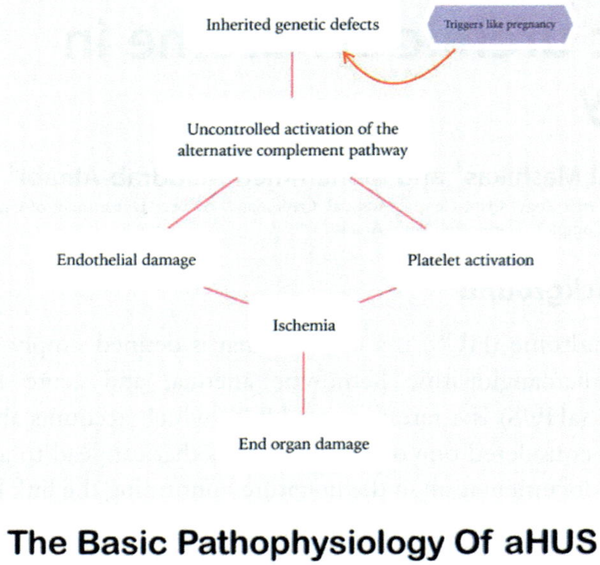

Figure 8.1 The basic pathophysiology of aHUS.

who is suspected of primary aHUS should be tested for pathogenic variants associated with it like CFH, MCP, CFI, C3, and CFB. In secondary aHUS, the recommendations regarding genetic testing for those patients are not clear, but posttransplant aHUS and pregnancy-associated aHUS are the exceptions and they should be tested [10].

The impact of acquired autoantibodies such as C3 nephritic factor and FH autoantibody must be followed in longitudinal studies to define their relevance to disease course in the context of the results of all other complement assays and genetic screening.

Pa-HUS has a clinical presentation similar to many serious pregnancy-related conditions. The most common diagnoses that can appear quite similar to pa-HUS are HELLP syndrome (hemolysis, elevated liver enzymes, and low platelet count syndrome), AFLP (acute fatty liver of pregnancy), and TTP (thrombotic thrombocytopenic purpura).

The first clue to reach this quite challenging diagnosis comes from the time of onset of the disease. HELLP syndrome, AFLP, and TTP usually happen in the third trimester, while pa-HUS usually appears postpartum [6].

HELLP syndrome and AFLP usually improve within 1 week after delivery, while patients with TTP and pa-HUS don't improve with delivery. HELLP syndrome is one of the severe features of preeclampsia, and is usually associated with hypertension and proteinuria. AFLP is usually associated with gastrointestinal symptoms like nausea and vomiting. TTP has unique neurological involvement. Pa-HUS has severe renal involvement.

DIC (disseminated intravascular coagulation) is rare in TTP and pa-HUS, but happens in around 50%—100% of patients with AFLP, and less than 20% of patients with HELLP syndrome. AKI could happen in any of them with mild degree, but it tends to be more moderate with AFLP, and is usually severe with pa-HUS.

Hemolytic anemia is usually present with HELLP syndrome and pa-HUS, but it is always present with TTP, while it could be absent with AFLP.

Partial thromboplastin time (PTT) is occasionally prolonged with AFLP, could be prolonged with HELLP syndrome, but it is usually normal with TTP and pa-HUS.

Hypoglycemia is pathognomonic for AFLP.

Thrombocytopenia is usually more prominent with TTP ($<20,000/mm^3$). It is usually less prominent in HELLP syndrome and pa-HUS ($>20,000/mm^3$), and it is usually better with AFLP ($>50,000/mm^3$).

Lactate dehydrogenate (LDH) is usually more than 600 IU/L in HELLP syndrome, but it is usually higher with TTP and pa-HUS (>1000 IU/L).

Ammonia could be elevated with AFLP, but it is usually normal with pa-HUS. Liver transaminase is always elevated with AFLP but not with pa-HUS. Von-Willebrand factor (VWF) multiverse could be present with TTP and pa-HUS but not with AFLP and HELLP. Deficiency of ADAMTS13 (a disintegrin and metalloproteinase with a thrombospondin type 1 motif, member 13) below 10% is always present with TTP, but not a must with pa-HUS. See Table 8.1.

Management

There are no clear guidelines to navigate the management of this difficult condition. The multidisciplinary team approach is key in the management of pa-HUS, as the collective knowledge of maternal—fetal medicine specialist, hematologist, intensivist, and nephrologist are needed.

The initial focus of the treatment is to replace the abnormal dysfunctional proteins and complements with normal proteins. This is why plasma exchange is the initial first step of the management.

There are some of the traditional treatments like plasma exchange and platelet therapy that have been used for many cases. Eculizumab is a promising drug which can effectively treat the condition. Controlling the blood pressure is a must to prevent further damage to the endothelium and the increased risk of end-organ damage.

Plasma exchange

Start plasma exchange as soon as possible after you obtain all the investigations needed, preferably during the first 24 hours of suspicion (e.g., renal and liver panel, PT, PTT,

Table 8.1 The clinical and laboratory differences between TTP, aHUS, AFLP, and HELLP syndrome.

	TTP	A-HUS	AFLP	HELLP syndrome
Time of onset	3rd trimester	Postpartum	3rd trimester	3rd trimester
Improve with delivery	No	No	Yes	Yes
Hypertension + proteinuria	—	—	—	Usually
Nausea/vomiting	—	—	Usually	—
Neurological involvement	Yes	—	—	—
Renal involvement	—	Yes	—	—
DIC	Rare	Rare	50%–100%	<20%
AKI	Mild	Severe	Moderate	Mild
Hemolytic anemia	Always	Usually	Could be absent	Usually
Prolonged PTT	Usually absent	Usually absent	Occasionally	Could be present
Hypoglycaemia	—	—	Always	—
Thrombocytopenia	<20,000/mm^3	>20,000/mm^3	>50,000/mm^3	>20,000/mm^3
LDH	>1000 IU//L	>1000 IU//L	—	>600 IU/L
Ammonia	—	Normal	Could be elevated	—
Liver transaminase elevation	—	No	Always	Always
VWF multiverse	Could be present	Could be present	No	No
ADAMTS13	Always	Could be present	—	—

CBC, blood glucose, LDH, ADAMTS13 level, ammonia level) [6]. This is because genetic testing, to diagnosis pa-HUS, will take a long time (up to a few weeks in some facilities).

Plasma exchange should be performed daily, and it is preferable to be pooled plasma from male donors to decrease the possible transfusion-related acute lung injury.

Plasma exchange will help to improve the platelet count but will not help to restore the kidney function. This is why the majority of patients will eventually have ESRD despite this treatment [2].

The success of plasma exchange comes in the form of normal platelet count and LDH value for two consecutive days, and then it can be stopped [6].

It is recommended to avoid transfusion of platelets even if platelets count is less than 50,000/mL especially if there is no bleeding, and will not need a procedure that can cause bleeding.

Renal replacement therapy

Renal replacement therapy can be considered if needed for signs of renal compromise, including hyperkalemia, uremia, metabolic acidosis, or oliguria that require urgent intervention to prevent further complications like pulmonary edema.

Blood transfusion

As with any other critical disease, we prefer to keep hemoglobin above 7 g/dL, by transfusing packed red blood cells as needed.

Eculizumab

It is an FDA-approved anti-C5 therapy that has been licensed to treat aHUS, if the diagnosis is confirmed. It causes terminal complement blockage and it may reverse the pathophysiological process of pa-HUS [7]. It is a recombinant monoclonal antibody that binds to C5 complement and prevents the subletting into C5a and C5b.

As opposed to plasma exchange, recent reports show that Eculizumab leads to an improvement in kidney functions, decreased need for kidney transplant, and improvement in quality of life, in addition to the improvement in platelet counts.

There are no safety data for using Eculizumab during pregnancy, but this is not an issue, as 85% of pa-HUS happens in postpartum period, which makes using it more feasible. The drug can cross the placenta as studies show the presence of Eculizumab in one third of cord samples but was not present in maternal breast milk.

Patients should receive meningococcal vaccine at least 2 weeks prior to starting Eculizumab. If you need to start the treatment in less than 2 weeks, then you can give antibacterial prophylaxis for 2 weeks to reduce the risk of meningococcal disease.

It is also recommended to give *Streptococcus pneumoniae* and *Hemophilus influenzae* vaccines.

Eculizumab can be started at an induction dose of 900 mg intravenously, once weekly, for four total doses, then a maintenance dose of 1200 mg in week 5, then 1200 mg every 2 weeks thereafter. There is no need to adjust the dose in case of renal or hepatic impairment. If the patient is simultaneously undergoing plasma exchange, she should receive a supplementary dose of 600 mg within 60 minutes after each exchange. There is no known optimal duration. Some experts recommend to continue Eculizumab for 6–12 months, including 3 months of stable kidney function [8].

Eculizumab is contraindicated in cases of unresolved ***Neisseria meningitides*** infections and hypersensitivity to the drug.

The most common side effects of Eculizumab are hypertension (59%), anemia (35%), leukopenia (24%), hypokalemia (18%), proteinuria (12%), and peripheral edema (8%).

Delivery

Delivery and termination of pregnancy do not improve pa-HUS and do not change its prognosis, and is usually reserved for obstetric reasons. Mode of delivery (vaginal delivery vs. Cesarean section) also does not change the prognosis. On the other hand, we have to keep in mind that termination of pregnancy will help to differentiate pa-HUS from other conditions that usually improve with delivery, like HELLP syndrome and AFLP.

Prognosis

In one study that included 87 patients with pa-HUS at the time of diagnosis, 71% of the patients needed dialysis, 78% underwent plasma exchange, 41% had plasma infusions, and 5% of them received Eculizumab [5].

During that study, with a follow up of 7.2 years, 53% had ESRD, while 91% had stable CKD (chronic kidney disease). 28% of patients had HUS relapse. 27% received kidney transplants. After which, recurrence rate reaches up to more than 50%.

In another study, 60%—70% of the patients with a-HUS will develop ESRD [2].

The type of genetic mutation does not affect prognosis [5].

Even if remission is achieved, the risk of relapse in subsequent pregnancies is high. That's why pregestational counseling is highly recommended.

Some reports show association between the oral contraceptive pills use and aHUS [9]. As other hormonal contraceptives methods were not studied and no recommendations is available, patient can use copper IUD as effective form of contraception till a recommendation is available [10].

References

[1] Noris M, Remuzzi G. Atypical hemolytic—uremic syndrome. N Engl J Med 2009;361(17):1676—87. https://doi.org/10.1056/NEJMra0902814.

[2] Fakhouri F, Roumenina L, Provot F, Sallée M, Caillard S, Couzi L, Essig M, Ribes D, Dragon-Durey MA, Bridoux F, Rondeau E, Frémeaux-Bacchi V. Pregnancy-associated hemolytic uremic syndrome revisited in the era of complement gene mutations. J Am Soc Nephrol 2010;21(5): 859—67. https://doi.org/10.1681/ASN.2009070706.

[3] Kavanagh D, Goodship TH. Atypical hemolytic uremic syndrome. Curr Opin Hematol 2010;17(5): 432—8. https://doi.org/10.1097/MOH.0b013e32833cae86.

[4] Bresin E, Rurali E, Caprioli J, Sanchez-Corral P, Fremeaux-Bacchi V, Rodriguez de Cordoba S, Pinto S, Goodship TH, Alberti M, Ribes D, Valoti E, Remuzzi G, Noris M. European Working Party on Complement Genetics in Renal Diseases. Combined complement gene mutations in atypical hemolytic uremic syndrome influence clinical phenotype. J Am Soc Nephrol 2013;24(3):475—86. https://doi.org/10.1681/ASN.2012090884.

[5] Bruel A, Kavanagh D, Noris M, Delmas Y, Wong EKS, Bresin E, Provôt F, Brocklebank V, Mele C, Remuzzi G, Loirat C, Frémeaux-Bacchi V, Fakhouri F. Hemolytic uremic syndrome in pregnancy and postpartum. Clin J Am Soc Nephrol 2017;12(8):1237—47. https://doi.org/10.2215/CJN.00280117.

[6] Saad AF, Roman J, Wyble A, Pacheco LD. Pregnancy-associated atypical hemolytic–uremic syndrome. AJP Rep 2016;6(1):e125–8. https://doi.org/10.1055/s-0036-1579539.

[7] Faguer S, Huart A, Frémeaux-Bacchi V, Ribes D, Chauveau D. Eculizumab and drug-induced haemolytic–uraemic syndrome. Clin Kidney J 2013;6(5):484–5. https://doi.org/10.1093/ckj/sft078.

[8] Goodship TH, Cook HT, Fakhouri F, Fervenza FC, Frémeaux-Bacchi V, Kavanagh D, Nester CM, Noris M, Pickering MC, Rodríguez de Córdoba S, Roumenina LT, Sethi S, Smith RJ. Conference participants. Atypical hemolytic uremic syndrome and C3 glomerulopathy: conclusions from a "kidney disease: improving global outcomes" (KDIGO) controversies conference. Kidney Int 2017;91(3): 539–51. https://doi.org/10.1016/j.kint.2016.10.005.

[9] Hauglustaine D, Van Damme B, Vanrenterghem Y, Michielsen P. Recurrent hemolytic uremic syndrome during oral contraception. Clin Nephrol 1981;15(3):148–53.

[10] Noris M, Bresin E, Mele C, Remuzzi G. Genetic atypical hemolytic–uremic syndrome [updated 2021 Sep 23]. In: Adam MP, Everman DB, Mirzaa GM, Pagon RA, Wallace SE, Bean LJH, Gripp KW, Amemiya A, editors. GeneReviews® [Internet]. Seattle (WA): University of Washington, Seattle; 2007. p. 1993–2023.

CHAPTER 9

Urinary tract infection and acute pyelonephritis in pregnancy

Vera Álava Fabricio[1], Moreira Macías Jéssica[2] and Rodríguez María Virginia[3]
[1]Manta General Hospital — IESS, Critical Care Ecuatorian Society (SECI), Laica Eloy Alfaro University of Manabí, Manabí, Ecuador; [2]Ecuatorian Pichincha Colposcopy Society, Manta General Hospital — IESS, Manabí, Ecuador; [3]Critical Care Ecuatorian Society (SECI), Rafael Rodríguez Zambrano Hospital — MSP, Laica Eloy Alfaro University of Manabí, Manabí, Ecuador

Definition

Bacterial, fungal, or viral infection of the urinary system (urethra, bladder, ureters, or kidneys) in a pregnant woman regardless of her gestational age.

In this chapter, we will review mainly bacterial infections due to their high frequency and severity, as well as the risks of complications associated with them.

Signs and symptoms

Urinary tract infection (UTI) in pregnant women may be asymptomatic to what is called in asymptomatic bacteriuria (ASB) or symptomatic that presents in order of frequency the following symptoms and signs: dysuria (pain or burning when urinating), pollakiuria (increased urinary frequency and in small amount), cloudy and stench urine (bad smell), pain in hypogastrium, pressure feeling in the bladder, fever, chills, low back pain, nausea, and vomiting. The presence of these symptoms and/or signs is related to the height of the infection, the most frequent being low infections (urethritis and cystitis).

Etiology

In most guidelines and publications such as Glaser's in 2015 [1], Small 2019 Cochrane reviews [2], Dube in 2022 [3], among others, emphasize the presentation of ASB and its progression to infection as such, being more frequently gram-negative bacteria such as *Escherichia coli* with a prevalence around 80%—90%, *Proteus mirabilis*, *Klebsiella pneumoniae*, *Enterococcus* spp., *Staphylococcus saprophyticus*, or *Streptococcus beta agalactiae* (GBS) in a smaller proportion.

The Kidney of the Critically Ill Pregnant Woman
ISBN 978-0-443-21473-8
https://doi.org/10.1016/B978-0-443-21473-8.00003-3

Risk factors

Pregnancy is associated as an important risk factor when compared with nonpregnant women and men, and taking only this population in this chapter, the main specific risk factors are the following [1]:

- History of previous urinary infections.
- Renal/ureteral lithiasis.
- Urogynecological malformations.
- Vesicoureteral reflux.
- Kidney failure.
- Diabetes mellitus.
- *Chlamydia trachomatis* infection.
- Multiparidad.

Types

When talking about UTIs in pregnant women, the same classification is made as in nonpregnant women according to the presence or absence of symptoms and then to the site of greatest compromise or onset of infection, having the following classification.

Asymptomatic bacteriuria

Diagnosis

Urine culture with >100,000 colony-forming units (CFUs) of bacteria per milliliter (mL), in two or more samples, without symptoms or clinical signs of UTI.

Since 1987 with Platt [4], in 2003 Nicolle [5], and recently in 2020, several Brazilian scientific societies make recommendations for the clinical management of asymptomatic lower UTIs in pregnant women due to their confirmed association with the development of symptomatic infections later during pregnancy (by 20%–30%), prematurity, and low birth weight [6], even Yan in 2018 reports the association with preeclampsia [7], and there are also studies with less statistical strength that associate it with other neonatal alterations.

Because it is recommended to perform cultures at least once during pregnancy, at the beginning of the second trimester, that is, between week 12 and 16 (U.S. Preventive Services Task Force, Infectious Diseases Society of America, American College of Obstetricians and Gynecologists, American Academy of Pediatrics, American Academy of Family Physicians, European Association of Urology, Scottish Intercollegiate Guidelines Network), preferably twice, and if ASB is demonstrated in pregnant women, it must be treated [8–10].

Treatment

The advantage that the diagnosis is made by microbiology allows to know the causative microorganism and its sensitivity, giving the opportunity to perform a targeted treatment. That is to say that with this information what should be determined is the antibiotic with the lowest risk to maternal and fetal health, the best strategy would be a short cycle of 4–7 days, even a single dose if the best option is fosfomycin trometamol (about this antibiotic, its use is being discussed a lot as it is a possible indication to treat multidrug-resistant germs and its use should be reserved for these conditions).

The first choice would be beta-lactams such as amoxicillin, cephalosporins of first and second generation, even third generation according to sensitivity and the existence of oral presentation, the next option is fosfomycin trometamol (with the consideration noted previously) if there is allergy to beta-lactams and allergy or resistance to fosfomycin would correspond nitrofurantoin. The following dosage is suggested:

- Amoxicillin, 500 mg, oral way (PO), every 8 h or every 6 h, for 4–7 days.
- Cefuroxime, 250 mg, PO, every 12 h, for 4–7 days.
- Cefixime 200 mg, PO, every 12 h, until completing 7 days.
- Other beta-lactams.
- Fosfomycin trometamol, 3 g, PO, single dose (separated from meals).
- Nitrofurantoin 50–100 mg, PO every 6 h, for 4–7 days.

Acute cystitis

Diagnosis

The presence of suggestive clinic with the symptoms and signs described allow a probable diagnosis, additionally positive urine dipstick can be found, with leukocytes, proteins (>1), red blood cells (>1), or nitrites, which in a symptomatic pregnant woman suggests infection, as well as the presence of bacteria in the urinary sediments, so antibiotic treatment will be started. However, urine culture should be performed to confirm the diagnosis and find resistance in the antibiogram. It is considered positive urine culture with >100,000 CFU/mL, with or without the presence of macroscopic/microscopic hematuria.

Treatment

Treatment is usually started empirically, as antibiogram is often not available. Therefore, the same scheme should be followed as in symptomatic bacteriuria.

Acute pyelonephritis

Diagnosis

In this case, you may or may not find the symptoms and signs of cystitis accompanied by fever and costovertebral pain with positive lumbar percussion fist and positive urine culture (>100,000 CFU/mL).

Treatment

To determine the place of treatment (ambulatory vs. hospital), the hemodynamic stability of the patient, the presence or absence of organic involvement (sepsis), sensitivity to enteral or parenteral presentation drugs, and of course the degree of fetal involvement or the presence of maternal/fetal complications must be evaluated.

Ambulant

In pregnant women without hemodynamic compromise, without evidence of organ failure, without fetal involvement, without comorbidities indicating maternal/fetal risk, treatment can be started on an outpatient basis.

Until the result of the cultures is available, empirical treatment should be initiated that should be based on local epidemiology (which is obtained through periodic reports of cultures of previous patients) either by place of residence or care of patients and should be complemented with the recommendations of the corresponding guidelines. In this situation, we maintain the scheme applied to ASB and cystitis, since the progression to pyelonephritis generally occurs in 20%–30% due to the progression of an untreated ASB, which can be reduced up to 3%–4% if treated before giving symptoms [11].

Optionally can be used at the time of care in the Health Unit Ceftriaxone, 1 g, intravenous (IV), single dose, and continue at home with Cefixime 200 mg, VO, every 12 h, until completing 7 days.

It is recommended that once the outpatient treatment is started, the patient is reassessed in 48–72 h, and it is defined if there is an adequate response or if she presents any criteria to be hospitalized, in addition it should be defined if it is necessary to modify the antibiotic spectrum according to the results of culture/antibiogram that should already be reported.

Hospital

The following criteria for hospital admission are: hemodynamic instability, sepsis (documentation of organ failure), dehydration and/or oral intolerance, threat of premature birth, persistent fever (≥38°C) and no clinical response after 72 h (3 days) of outpatient treatment, recurrent pyelonephritis and allergy to beta-lactams, gestational ≥ to 24 weeks with risk factors (comorbidities); at least 3 days of parenteral treatment should be completed and afebrile for 48–72 h maintained to define the switch to enteral medication and subsequent outpatient treatment [3].

Empirical use of antibiotic will be assessed for the risk factors to colonization of *E. coli* producers of extended-spectrum beta-lactamases (ESBL), such as: comorbidities (chronic renal failure, diabetes mellitus, liver cirrhosis, chronic obstructive pulmonary disease), immunosuppression, urinary catheter user, antecedent of admission greater than 48 h in the previous 3 months, stay in intensive care unit (ICU), recent surgical interventions, and systemic antibiotic use in the previous 3 months. Another consideration is infection in the puerperium that can be associated with Enterococcus and merits the use of aminopenicillins in combination.

Suggested empirical treatments:

Pregnant:

No ESBL risk factors:

- First choice: Ceftriaxone 1 g, IV, every 24 h.
- Second choice: Amoxicillin 1g + clavulanic acid 200 mg, IV, every 8 h or Ampicillin 1 g + Sulbactam 500 mg, IV, every 6 h.
- Allergy to beta-lactams: Aztreonam 1 g, IV, every 8 h.

With ESBL risk factors:

- First choice: Ertapenem 1 g, IV, every 24 h.
- Allergy to beta-lactams: Fosfomycin 4 g, IV, every 8 h.
- Optional in the second trimester risk/benefit: Amikacin 1 g, IV or intramuscular (IM), every 24 h.

Puerperal:

No ESBL risk factors:

- First choice: Ceftriaxone 1 g, IV, every 24 h + Ampicillin 2 g, IV, every 6 h.
- Second choice: Piperacillin 4 g + Tazobactam 500 mg, IV, every 6 h (monotherapy).
- Allergy to beta-lactams: Aztreonam, 1 g, IV, every 8 h + Teicoplanin 600 mg, IV, every 24 h.

With ESBL risk factors:

- First choice: Ertapenem 1 g, IV, every 24 h + Ampicillin 2 g, IV, every 6 h.
- Allergy to beta-lactams: Fosfomycin 4 g, IV, every 8 h + Teicoplanin, 400 mg IV, every 12 h (three doses) followed by 400 mg, IV, every 24 h.

The previous results of urine cultures and their antibiograms should be monitored because the presence of ASB is a risk factor for pyelonephritis and when using empirical therapy, the antibiotics previously used should be considered, evaluating the infectious agent and its sensitivity to antibiotics to adjust the therapy according to the results of the antibiogram.

In case of rule out the ESBL germ, it is recommended to aim the antibiotic treatment according to the results of the culture and antibiogram. If there is no response, the possibility of more complex treatments should be considered in consultation with the infectious disease service.

Medical monitoring

In the case of ASB in pregnancy, resolution should be validated with a urine culture after 7–15 days after completing treatment, if there is recurrence, act according to the result of the new antibiogram. In addition, it is advisable to do a urine culture every 3 months, if it is not feasible, perform it at least in the third trimester.

In cases of UTIs in pregnant women (symptomatic), it is recommended to perform periodic controls with cultures and antibiograms to rule out ASB or recurrences, at least in the third trimester.

Special cases

Prophylaxis of recurrent UTI

Includes antibiotic treatment to prevent recurrence of UTI, recommended in cases of ASB or recurrent cystitis (three or more positive urine cultures), despite having received complete effective treatment, after the third treatment 4–7 days and after recurrent pyelonephritis during pregnancy or a positive urine culture properly treated.

Prophylaxis should be considered after a single episode of pyelonephritis in patients with renal or excretory tract pathologies, as well as in cases of immunosuppression or other serious maternal medical conditions.

Antibiotic prophylaxis is carried out after the treatment regimen is completed, and continues until 4–6 weeks after delivery. The recommended drugs are:

- First choice: Cefalexin 250 mg, PO, every 24 h (preferably administered in the evening).
- Second choice: Fosfomycin trometamol 3 g, PO, every 5 days.

Asymptomatic bacteriuria due to GBS

The presence of GBS in the urine signals high colonization in the genital tract and is linked to an increased risk of kidney infections, infections in the placenta, and early neonatal sepsis.

If more than 100,000 CFU of GBS is found in the urine, antibiotic treatment is recommended for ASB, but with 10,000–100,000 CFU, initiation of treatment is not advised, as most women recolonize rapidly. In case of contamination, the urine test will be repeated.

Prophylaxis during delivery for GBS is recommended in both cases, but a vaginal–rectal culture is not necessary between weeks 35 and 37.

Prophylaxis during childbirth:

- First choice: Penicillin G 5 million IU, IV, initial dose 4 h before delivery, then 2.5 million, IV, every 4 h until the end of labor.
- Second choice: Ampicillin 2 g, IV, initial dose 4 h before delivery, then 1 g, IV, every 4 h until the end of labor.
- Allergy to beta-lactams: Clindamycin 900 mg, IV, every 8 h until the end of labor or Vancomycin 1 g, IV, every 12 h (or 20 mg/kg every 8 h, maximum dose 4 g/day) until the end of labor.

Nosocomial febrile urinary infection

Urinary infection that appears after 48 h of hospital admission, that is, with a high suspicion that the causative microorganism was acquired in the hospital, so it is important to know the hospital epidemiology by performing and updating every 6 months of a microbiological map, which allows to know the bacteria and fungi prevalent in the different hospital areas allowing in this way to select the best possible coverage of according to the location of the pregnant woman (ICU, obstetric center, delivery room, obstetric hospitalization). The recommended drugs, if microbiological information is not available, are:

- First choice: Piperacillin 4 g + Tazobactam 500 mg, IV, every 6 h + Teicoplanin 400 mg, IV, every 12 h (three doses) followed by 400 mg, EV, every 24 h.
- Second choice: Meropenem 1 g, IV, every 8 h in extended infusion + Teicoplanin 400 mg, IV, every 12 h (three doses) followed by 400 mg, IV, every 24 h.
- Allergy to beta-lactams: Fosfomycin 4 g, IV, every 8 h + Teicoplanin 400 mg, IV, every 12 h (three doses) followed by 400 mg, EV, every 24 h.

* Vancomycin can be used as an alternative to Teicoplanin in specific situations or under the criteria of infectiology.

Fungal infections in pregnancy urinary tract

Although the presence of fungal infections in pregnant women has not been widely studied, colonization of the urogenital tract by fungi, in a greater proportion candida species, is very frequent and is related to recurrent UTIs. As described by Ghaddar in a 2020 review in Lebanon, the presence of Candida in the urogenital tract is related to gestational diabetes in the case of *Candida albicans* and to preterm delivery and other gestational complications in the presence of *Candida glabrata* and *Candida krusei*.

Most of the current guidelines do not recommend the treatment of asymptomatic candidiasis, even so it should be performed in case of candidemia, specifically in the presence of organ failure (cardiovascular or other). The possibility of congenital disorders is more related to the azole therapy than to amphotericin B deoxycholate, which is why its use is recommended over azoles. In addition, in the reports of resistance, less resistance was found in amphotericin B. In regarding echinocandins, there is still little information in this regard, but caspofungin is related to intrauterine ossification, so the risk-benefit must be weighed.

Goods practices

- Assess the medical history, make an adequate physical examination, evaluate the results of a previous urine culture, if there was a previous UTI and if it was treated correctly. In addition, general laboratory tests (blood count, electrolytes, creatinine, and C-reactive protein) and a urinary sediment will be performed.
- Between weeks 12 and 16 of gestation, ASB should be detected and treatment administered to reduce the risk of pyelonephritis and maternal and fetal complications.
- Recommend to the pregnant woman modify the lifestyle, drink enough fluids (more than 2000 mL every 24 h), urinate frequently, and completely after having sex, reduce sexual contact with several partners and condoms use.
- Inform the patient about hygienic measures to prevent reinfection when receiving antibiotic treatment. These include avoiding immersion baths, washing hands before

using the bathroom, using liquid soap, wearing cotton underwear, wiping from front to back after urinating or defecating, and starting genital and perineal hygiene for the urethral meatus.

- Urinalysis should be taken following proper hygiene measures, such as cleaning the external genitalia and collecting mid-flow urine. Bladder catheterization is only necessary in special cases that may affect the interpretation of the result, such as the presence of blood in recent postpartum women.
- A contaminated urine culture is identified when between 10,000 and 100,000 CFUs per milliliter are found or when three or more germs are present. If the urine culture is contaminated, it should be repeated following appropriate asepsis measures. If there are three consecutive crops that are contaminated, it is considered negative.
- The selection of the antibiotic must be safe for both mother and fetus, since many drugs pass through the placenta, the choice should be based on the susceptibility of the patient, the resistance documented in the area and the availability of drugs.
- Perform renal ultrasound in risk situations, such as lack of response to antibiotic treatment in 48–72 h, recurrent ASB episodes, cystitis or pyelonephritis, involvement of the general condition, symptoms compatible with abscess, sepsis, presence of hematuria and even pyelography if the ultrasound is inconclusive and there is no response to treatment.

References

[1] Glaser AP. Urinary tract infection and bacteriuria in pregnancy. Urol Clin N Am 2015;42(4):547–60. https://doi.org/10.1016/j.ucl.2015.05.004.

[2] Small FM, Vazquez JC. Antibiotics for asymptomatic bacteriuria in pregnancy. Cochrane Database Syst Rev 2019;2019(11):CD000490. https://doi.org/10.1002/14651858.CD000490.pub4.

[3] Dube R, Al-Zuheiri STS, Syed M, Harilal L, Zuhaira DAL, Kar SS. Prevalence, bacteriological clinical profile and antibiotic resistance of symptomatic urinary tract infections in pregnant women. Antibiotics (Basel) 2023;12(1):33. https://doi.org/10.3390/antibiotics12010033.

[4] Platt R. Adverse consequences of asymptomatic urinary tract infections in adults. Am J Med 1987; 82(6B):47–52. https://doi.org/10.1016/0002-9343(87)90618-8.

[5] Nicolle LE, Bradley S, Colgan R, et al. Infectious Diseases Society of America guidelines for the diagnosis and treatment of asymptomatic bacteriuria in adults. Clin Infect Dis 2005;40(5):643–54. https://doi.org/10.1086/427507.

[6] de Rossi P, Cimerman S, Truzzi JC, Cunha CAD, Mattar R, Martino MDV, Hachul M, Andriolo A, Vasconcelos Neto JA, Pereira-Correia JA, Machado AMO, Gales AC. Joint report of SBI (Brazilian Society of Infectious Diseases), FEBRASGO (Brazilian Federation of Gynecology and Obstetrics Associations), SBU (Brazilian Society of Urology) and SBPC/ML (Brazilian Society of Clinical Pathology/Laboratory Medicine): recommendations for the clinical management of lower urinary tract infections in pregnant and non-pregnant women. Braz J Infect Dis 2020;24(2):110–9. https://doi.org/10.1016/j.bjid.2020.04.002.

[7] Yang Y, Le Ray I, Zhu J, Zhang J, Hua J, Reilly M. Preeclampsia prevalence, risk factors, and pregnancy outcomes in Sweden and China. JAMA Netw Open 2021;4(5):e218401. https://doi.org/10.1001/jamanetworkopen.2021.8401.

[8] Lin K, Fajardo K, U.S. Preventive Services Task Force. Screening for asymptomatic bacteriuria in adults: evidence for the U.S. Preventive Services Task Force reaffirmation recommendation

statement. Ann Intern Med 2008;149(1):W20–4. https://doi.org/10.7326/0003-4819-149-1-200807010-00009-w1.
[9] Ressel G. Introduction to AAFP summary of recommendations for periodic health examinations. American Academy of Family Physicians. Am Fam Physician 2002;65(7):1467.
[10] AAP Committee on Fetus and Newborn, ACOG Committee on Obstetric Practice. Guidelines for perinatal care. 7th ed. Elk Grove Village (IL): American Academy of Pediatrics; 2012.
[11] Gilstrap LC 3rd, Ramin SM. Urinary tract infections during pregnancy. Obstet Gynecol Clin N Am 2001;28(3):581–91. https://doi.org/10.1016/s0889-8545(05)70219-9.

Further reading

[1] Instituto Mexicano del Seguro Social. Prevención, diagnóstico y tratamiento de la infección del tracto urinario bajo durante el embarazo en el primer nivel de atención, México; 03/11/2016. Disponible en: http://imss.gob.mx/profesionales-salud/gpc http://www.cenetec.salud.gob.mx/contenidos/gpc/catalogoMaestroGPC.html.
[2] Habak PJ, Griggs Jr RP. Urinary tract infection in pregnancy [Updated 2022 Jul 5]. In: StatPearls [Internet]. Treasure Island (FL): StatPearls Publishing; 2023. Available from: https://www.ncbi.nlm.nih.gov/books/NBK537047/.
[3] Hospital Clínic, Hospital Sant Joan de Déu, Universitat de Barcelona. Protocol urinary tract infection and gestation. 2018. https://portal.medicinafetalbarcelona.org/protocolos/es/patologia-materna-obstetrica/infecciones-urinarias-y-gestacion.html.

CHAPTER 10

Lupus nephritis and pregnancy

Fatemeh Masjedi and Jamshid Roozbeh
Shiraz Nephro-Urology Research Center, Shiraz University of Medical Sciences, Shiraz, Iran

Introduction

Systemic lupus erythematosus (SLE) is a complex and chronic autoimmune disease characterized by relapsing systemic inflammation, whose exact cause remains unclear. The disorder has consistently been identified as a significant diagnosis in women, characterized by a multisystem pathology often seen at the childbearing age [1]. Around 4500 pregnancies annually in the United States are reported among women diagnosed with SLE [2]. The prevalence of renal disease in individuals with SLE is high, with about 50% of patients experiencing clinically detectable renal damage throughout the progression of their disease [3].

A unique physiological stress experienced by women during pregnancy and the systemic changes that occur during pregnancy may impact the progression of SLE and its renal symptoms. During the gestational period, fluctuations in sex hormones induce modifications in the immune system, facilitating the fetus's growth and inhibiting the mother's rejection of fetal antigens [4]. These modifications may potentially lead to an intensification of SLE activity. Despite contradictory clinical research, the prevailing data indicate that pregnancy is associated with an elevated risk of SLE flare and heightened disease activity [5,6]. The physiological changes associated with pregnancy might also result in renal stress. The augmentation of blood volume and glomerular filtration rate (GFR) imposes a further burden on the kidney, potentially leading to a decline in renal function in females with preexisting renal disorders [7,8].

Lupus nephritis (LN) may manifest during pregnancy as a first occurrence, while women with a prior history of LN have an elevated likelihood of experiencing disease flares during pregnancy. Women diagnosed with LN are also susceptible to experiencing negative maternal and fetal outcomes, including hypertension, fetal mortality, preterm birth, and intrauterine growth restriction (IUGR) [9,10]. The worst pregnancy outcomes are seen when the illness has shown activity during the 6 months before conception [11].

To ensure optimized results, healthcare providers must educate patients early about the risks associated with pregnancy during active disease, teratogenic effects of drugs, and contraception and family planning, where applicable. Ultimately, successful pregnancies are possible despite the risks associated with pregnancy in patients with LN, under the medical support of a multidisciplinary team, including obstetricians, rheumatologists, and nephrologists. Over the last several decades, there has been a significant improvement

The Kidney of the Critically Ill Pregnant Woman
ISBN 978-0-443-21473-8
https://doi.org/10.1016/B978-0-443-21473-8.00005-7

in maternal and fetal outcomes. The observed enhancement may be attributed to many variables, including the implementation of more efficacious therapy regimens that facilitate the achievement of SLE remission, the strategic timing of pregnancies during disease inactivity, and the meticulous monitoring and supervision during pregnancy [12].

In this chapter, the authors conducted a comprehensive examination of previously published original research, systematic reviews, metaanalyses, guidelines, recommendations on monitoring and therapeutic strategies, as well as risk assessment for unfavorable pregnancy outcomes associated with LN to provide an agreement on the LN management during pregnancy by utilizing evidence-based approaches.

Main changes in immune response strategy during pregnancy

Pregnancy is widely recognized as a distinct physiological event that presents a particular challenge to immune reactions. The female immune system employs a sophisticated approach to maintain immune quiescence during this period. It involves facilitating maternal immune tolerance toward fetal antigens while simultaneously preserving the capacity to defend against external pathogens [13,14].

During the first trimester, the immune system at the local level undergoes a proinflammatory phase characterized by the involvement of several immune cells, including macrophages, dendritic cells, neutrophils, and natural killer (NK) cells. These cells work together in a regulated manner to mount an inflammatory response. The proinflammatory response must control various activities for effective implantation and placentation, including trophoblast invasion, proliferation, tissue remodeling, and angiogenesis. There is a noticeable shift toward an immunologically tolerant condition throughout the second trimester, defined by the predominance of antiinflammatory Th2 responses. Regression of the Th1 and Th17 arms coincides with this transition, promoting fetal allograft tolerance and protecting the trophoblasts from future infections. In order to aid myometrial contraction and labor, the proinflammatory state rises again in the third trimester [15,16].

The T regulatory cells, also known as T regs, play a crucial role in maintaining immunological homeostasis and reducing oxidative stress during pregnancy. Throughout pregnancy, there is a concurrent rise in the population of T regs. This increase facilitates initiating an immune regulatory mechanism, leading to an enhanced proinflammatory response to fetal antigens (allograft). This phenomenon is particularly prominent during the first and second trimesters, as it aims to suppress the unfavorable activities of the prevailing Th1, Th2, and Th17 cells. In pregnant individuals with SLE, there is a concurrent occurrence of impaired regulatory T-cell function and an increase in the Th17 cytokine environment, which is influenced by estrogen. This leads to greater production of IL-17, which does not decrease with progesterone elevation, as seen in typical pregnancies [17–19]. The factor mentioned above has been associated with the etiology of unfavorable maternal and fetal consequences, such as hypertension, recurrent miscarriages, and premature labor. Additionally, it has been linked to impaired immunological responses to infections occurring during pregnancy [20,21].

Preconception risk analysis in SLE-affected women

The complex and dysregulated immunological characteristics of lupus, along with the existence of uncontrolled illness, pose an elevated risk for pregnancy complications. The ability to conceive in individuals with lupus is influenced by several factors, including the current state of the illness, the profile of autoantibodies present, and any other medical conditions that may be present. Consequently, it is recommended that pregnancies in individuals with SLE be pursued only when the disease has been inactive for a minimum of 6 months before conception to ensure a healthy result. This proposal strongly advocates counseling services for planned pregnancies [22,23].

Before choosing to conceive a female with lupus, it is essential to do a thorough evaluation to guarantee the well-being and safety of the mother and fetus. Suggestions for appropriate evaluation have encompassed personalized risk assessment by examining the current and past disease history, accurately characterizing the disease phenotype in each subject, assessing disease activity state, analyzing serological variables, considering the recurrence rate of disease flares, taking into account the presence of comorbidities related or unrelated to the systemic disease, evaluating the organ-specific involvement, assessing the possible risks associated with teratogenic drugs used to manage disease activity when necessary, and considering the range of safe and effective therapeutic options that are available. The evaluation should be accompanied by a personalized categorization of pregnancy-related risks and a comprehensive strategy for addressing potential important clinical scenarios in instances of exacerbation during pregnancy, using a tailored methodology. Counseling about contraception is recommended for women with uncontrolled and recurrent flares, severe organ failure, or comorbidity [24,25].

The evaluation of conception strategy for a pregnant woman with lupus should be based on a personalized approach to preconception requirements, considering the following clinical deliberations: Table 10.1 presents a comprehensive overview of the critical evaluation requirements that should be addressed before conception [26].

LN-associated maternal and fetal complications during pregnancy

Maternal morbidity and mortality incidence is much higher in individuals diagnosed with SLE, and the occurrence of SLE during pregnancy is linked to an elevated risk of morbidity [27]. SLE is often associated with many consequences, including neurologic disorders, nephritis, and thrombocytopenia. Certain individuals diagnosed with SLE

Table 10.1 Essential personalized preconception evaluation requirements in pregnancy with lupus [26].

Previous history of obstetrics and gynecology
Current disease activity state and in the last 6 months
Clinical recurrence rate of flares
Related and unrelated comorbidities
Careful consideration of the possibility of acute or chronic reduction of renal function in women with lupus nephritis
Serological variables (including positive antiphospholipid, anti-Ro, anti-La, antiU1 RNP antibodies)
Disease damage levels
Drug history relative to the disease condition
Current drug regimen and dose
Classification of individual risk
Selecting available individualized treatment options for managing flares during pregnancy
Education and counseling of patients and their families

may also exhibit antiphospholipid syndrome (APS), a condition linked to a greater risk of experiencing abortion and thrombosis [28].

In a national study conducted in the United States, including a large sample size of over 16 million pregnancies, it was shown that individuals diagnosed with SLE had a significantly elevated risk of experiencing thrombocytopenia, thrombosis, multiorgan damage, infection, and requiring blood transfusion, in comparison to those without SLE. Furthermore, a substantial twenty-fold rise in maternal mortality rate was also reported [29].

The incidence of active LN during pregnancy ranges from 4% to 30% [30]. The previous metaanalysis revealed a substantial association between active nephritis during pregnancy and poor outcomes, such as hypertension and premature birth [5]. There is a potential correlation between a history of LN and an increased likelihood of experiencing a disease flare [9,31]. A study including 113 pregnancies in 81 women with LN diagnosed before conception revealed the occurrence of 34 flares (30%), of which 20 were reversible by therapeutic interventions. Patients who had a prepregnancy renal function of either more than 1 g protein/24 hours or a GFR less than 60 mL/minutes/body surface area exhibited a nine-fold higher chance of experiencing a renal flare. Furthermore, low-dose aspirin therapy and normocomplementemia at conception were associated with a decreased risk of fetal loss [9].

The preeclampsia risk varies between 10% and 35% in women with SLE [30,32]. Women with underlying chronic kidney disease (CKD) and active LN are at predominantly higher risks [5,33]. Individuals with active disease at the time of conception, renal involvement, persistent hypertension, high-dose prednisone usage, or aPL antibodies are at the highest risk of preeclampsia [34,35]. Furthermore, women with LN before pregnancy are more likely to develop preterm preeclampsia and are at a higher risk of premature birth [10].

In conjunction with APS, SLE is correlated with a three-fold increase in the danger of pregnancy loss [36]. The most significant risk factors associated with abortion are the simultaneous presence of antiphospholipid (aPL) antibodies and active kidney problems [36–38].

Fetal growth restriction (FGR) is common in SLE-affected pregnancies. Although accurate statistics are missing, the risk ranges between 6% and 35%. Another study demonstrated that those with mild sickness were more susceptible to FGR, even after controlling for confounding variables, including hypertension and renal impairment [34].

SLE is also linked to an elevated likelihood of preterm delivery, perhaps attributable to the heightened occurrence of preeclampsia and FGR [39]. The risk factors for preterm delivery are similar to those contributing to preeclampsia, pregnancy loss, and FGR. They include increased disease activity, nephritis, chronic hypertension, and aPL antibodies [40].

In a recent study, Attia et al. investigated whether inactive LN at the onset of conception carries an increased risk of pregnancy complications compared to pregnancies without a history of LN. One hundred nineteen pregnancies were studied: 72 in group I (with a history of LN) and 47 in group II (nonrenal SLE). At conception, 16 (22.2%) of the renal pregnancies experienced renal flares. Maternal problems, notably renal flares, were recorded in 36 (50%) of group I and 13 (27.7%) of group II pregnancies, with a significant difference ($P = .015$). There were no significant variations in the frequency of pregnancy-related maternal and fetal problems between the two groups [41]. Therefore, LN per se is not a risk factor for poor pregnancy outcome; it is the LN activity at the pregnancy onset [41,42].

In another study, the pregnancy outcomes of 158 pregnancies were retrospectively investigated in 155 patients with LN, and for the first time, demonstrated that LN patients with a complete renal remission (CRR) duration greater than 18 months were related to good pregnancy outcomes and lower LN relapse. It also revealed that positive anti-C1q antibody before pregnancy is an independent risk factor for LN relapse; positive aPL antibody and prednisone dosage $\geq$12.5 mg/day during pregnancy are independent risk factors for fetal loss and premature birth, respectively; and age >30 years is an independent risk factor for preeclampsia and premature birth [43].

A meta-analysis showed an increased risk of premature birth associated with active LN [5]. A cohort study found that 16% (27 of 166) of SLE pregnancies finished with fetal loss, and the risk was 2.6 times greater in women who had proteinuria in the first trimester, indicating underlying renal involvement [44].

The type of LN may also affect fetal outcomes. In a work comparing outcomes in women with WHO class III/IV disease and WHO class II/V disease against women with SLE and no signs of nephritis, women with class III/IV disease had a considerably lower mean birthweight than the other two groups [45]. Larger multicenter studies may be able to address the question of whether the WHO histologic class is related to pregnancy outcomes.

Moreover, 33.3% of pregnancies showed persisting LN one year after delivery. The median postpartum follow-up duration was 5.9 (3.1−9.7) years, and CKD developed in 21.4% of pregnancies with renal flare. Failure to attain a 50% reduction in urine protein levels within 6 months, a longer overall duration of renal flare, and acute kidney injury (AKI) at renal flare were all related to CKD progression in individuals with renal flare. When patients undergo renal flares during pregnancy, it is critical to reduce proteinuria by more than 50% within 6 months and establish early remission to ensure favorable long-term renal outcomes [46].

Despite the increased risk of unfavorable maternal and fetal outcomes in women with SLE, particularly those with LN, data suggest that outcomes have improved in recent decades. Clark et al. found that the rate of fetal loss has decreased from 43% in the 1960s to a mean of 17% from 2000 to 2003, which is consistent with the rate of pregnancy loss in the general population, and there may also be a decrease in preterm births in SLE pregnancies [47]. This improvement might be attributed to improvements in SLE therapy, better counseling and contraceptive usage, and modifications in prenatal care.

Overall, LN increased the odds of unfavorable pregnancy outcomes and drug usage in SLE patients. Pregnant women with LN should get extra care and be closely monitored [48].

Evidence-based consensus on drug management in pregnancy with LN

In actual practice, it is very important to consider how various therapy approaches affect fertility and conception in LN patients. When deciding whether to employ a certain management technique, weighing the risk versus benefit is important. The prenatal counseling, natal, and postnatal timeframes are all covered under the risk to benefit policy.

Blood pressure drugs: To avoid disease progression or other end-organ damage, blood pressure should be controlled to a target of systolic blood pressure less than 140 mm Hg and diastolic blood pressure less than 90 mm Hg in women with underlying renal disease. Because blood pressure fluctuates during pregnancy, it should be constantly monitored, and drugs should be titrated [49]. Angiotensin-converting enzyme inhibitors and angiotensin-receptor blockers should be stopped before pregnancy due to the risk of fetal abnormalities [50,51]. Other medications to avoid during pregnancy include atenolol, which has been linked to IUGR, and spironolactone, which can cross the placenta and have antiandrogen effects on the fetus. Methyldopa, hydralazine, nifedipine, labetalol, and diuretics are all considered safe during pregnancy [52].

Anticoagulation and aspirin: Given the risk of thrombosis, women with nephrotic range proteinuria should be evaluated for anticoagulation. Due to the higher risk of preeclampsia in SLE patients, European guidelines advocate low-dose aspirin treatment for all patients [11]. There is also evidence that aspirin therapy in SLE patients may result in better prenatal outcomes, including a lower chance of fetal loss [9]. Furthermore, due to the risk of developing preeclampsia, individuals with a history of active LN should be on low-dose aspirin (100 mg/day) before week 12 [53].

Immunosuppressive drugs: Immunosuppressive treatment for SLE is appropriate during pregnancy in two scenarios: induction therapy for a new diagnosis of SLE or a flare of existing illness or maintenance therapy for individuals in remission. However, therapeutic options are restricted since typical LN medicines, such as cyclophosphamide and mycophenolate mofetil (MMF), are teratogenic. As a result, corticosteroids, hydroxychloroquine, azathioprine, and calcineurin inhibitors are the cornerstones of LN treatment in pregnancy, with less data available on biologic agents such as rituximab and belimumab.

Steroids: Corticosteroids have long been used to treat LN and are effective. However, they are rarely used as monotherapy, except for lupus podocytopathy and minimal change disease (MCD) associated with SLE [11]. Steroids have some adverse effects in general, including fluid retention, hyperglycemia, and an increased risk of infection, and have been linked to gestational diabetes, sepsis, and premature birth [54]. Despite these concerns, steroids are often used in both induction and maintenance therapy during pregnancy, and their safety profile is superior to that of other recognized teratogenic drugs. A strategy of starting an additional medication, such as a calcineurin inhibitor, to reduce steroid exposure is beneficial, and remission can be accomplished [55]. There is no evidence that steroids would prevent an SLE flare during pregnancy; thus, they should not be used prophylactically [6].

According to the currently accepted evidence-based, using nonfluorinated corticosteroids, including prednisolone, hydrocortisone, and methylprednisolone, are recommended as safe options for maintaining disease quiescence during pregnancy [56].

Hydroxychloroquine (HCQ): HCQ is an antimalarial medication frequently used to treat SLE during pregnancy. There have been no reports of birth abnormalities, and despite modest levels of transfer through breast milk, it is likely safe in breastfeeding [57].

A recent metaanalysis of 119 pregnancies showed that the risk of preeclampsia, intrauterine distress, preterm birth, and postpartum hemorrhage was not increased by HCQ [25]. Another metaanalysis on 938 pregnancies in 804 women also revealed that women who continued using HCQ during pregnancy with SLE had less disease activity and healthy pregnancies [58]. According to the data about the relationship between medication and pregnancy-related problems in SLE, HCQ may reduce the incidence of preeclampsia, spontaneous abortions, and congenital abnormalities [59–61].

There were no negative fetal outcomes associated with HCQ exposure in a prospective study on three groups of SLE pregnancies—HCQ used throughout pregnancy

($n = 56$), no HCQ therapy ($n = 163$), and cessation of HCQ therapy up to the first trimester ($n = 38$). However, patients who discontinued therapy had a higher risk of flare and increased disease activity [62]. In pregnancies complicated with LN, it is currently advised to utilize HCQ as the first-line maintenance therapy [63].

Antimetabolites: Due primarily to its usage in kidney transplantation, azathioprine has a long history of being safe during pregnancy [64]. To maintain remission throughout pregnancy, azathioprine may be used instead of MMF treatment [11]. Azathioprine has been widely used in the context of pregnancy. Adverse reactions are uncommon, and there have been no reports of teratogenicity in humans [64,65]. In situations of severe SLE flare, it is generally believed that the advantages exceed the risks. Azathioprine has been identified as the preferred immunosuppressant for the maintenance therapy of individuals with preexisting LN and the induction of newly diagnosed cases [66].

Calcineurin inhibitors: Calcineurin inhibitors (cyclosporine and tacrolimus) have also proven safe during pregnancy. Cyclosporine is a good option for lupus flares that are resistant to conventional medicinal treatments. Proliferative LN may be treated with it effectively [67]. Tacrolimus, a calcineurin inhibitor, has been shown to be more effective than cyclosporine at inducing remission in treating LN [68]. However, it has been seen that tacrolimus is linked to a higher likelihood of developing preeclampsia, gestational diabetes, and hypertension when compared to cyclosporine [2]. The efficacy of tacrolimus in inducing remission of LN during pregnancy has been demonstrated, although studies have been limited to case reports and case series [55,69].

Nevertheless, in a new clinical practice, tacrolimus was suggested as a valid treatment option for LN in fertile women in Japan, and with careful monitoring, pregnant women with LN could continue their tacrolimus treatment without unexpected adverse events/safety concerns [70]. Yet, due to the increased volume of distribution observed during pregnancy, it is advisable to initiate calcineurin inhibitors at a minimally effective dosage and regularly monitor serum levels to prevent potential toxicity [66].

Cyclophosphamide: Cyclophosphamide is a recognized teratogen, and miscarriage and fetal death have been documented in SLE patients exposed to it during pregnancy [71]. However, literature from breast cancer patients shows that using cyclophosphamide in the second and third trimesters can result in favorable fetal outcomes [72].

Furthermore, there are case reports from women with SLE and LN being effectively treated with cyclophosphamide during pregnancy [73,74]. Cyclophosphamide should not be utilized as a first-line therapy for LN in pregnancy due to a lack of evidence for safety in this group and the availability of alternate medications.

Biologic agents: New biologic medications such as belimumab, infliximab, rituximab, golimumab, and adalimumab have been utilized for various autoimmune illnesses, including SLE, during the last several years. Although there is a lack of information on the prenatal effects of these drugs, early exposure during pregnancy and the use of these drugs in patients who have failed to respond to other treatments are providing additional data [75].

Initially showing promise for treating LN, rituximab is a monoclonal antibody that depletes B cells. However, later randomized studies have not shown a substantial effect [76]. According to current evidence, using rituximab in the first trimester of pregnancy does not raise the risk of fetal abnormalities. Despite this, there is a high risk of B-cell depletion and cytopenias in newborns, which increases the risk of infectious events. As a result, rituximab should be administered only in rare cases [77,78].

Belimumab, a monoclonal antibody against soluble B lymphocyte stimulator, has been approved by the U.S. Food and Drug Administration (FDA) for the treatment of both SLE and LN after being demonstrated to be effective in clinical trials for individuals with both conditions [79,80]. A case with refractory LN during pregnancy was treated with belimumab, and the patient's disease was controlled with this therapy. However, the fetus was born with a mild Ebstein's anomaly [81].

Other biologics are being developed; however, due to the lack of safety information about their use during pregnancy, avoiding these more recent biologic drugs is typically recommended until all other options with superior safety profiles have failed and patients have been informed of any possible unknown challenges.

Intravenous immunoglobulin (IVIG): Although not first-line treatment, IVIG may treat active lupus flares or nephritis that have not responded to corticosteroids, azathioprine, or HCQ [66,82]. Since LN may be fatal, the advantages of these strong medications often exceed any potential risks. A patient with diffuse proliferative LN was treated in pregnancy with a combination of IVIG, etanercept, and plasmapheresis, with favorable maternal and fetal outcomes [83].

Conclusion

Several immunologic and physiologic alterations during a normal pregnancy can be problematic for a patient with SLE. Women with LN are at increased risk for adverse maternal and neonatal outcomes, such as preeclampsia, fetal loss, preterm delivery, and IUGR. A multidisciplinary team should thoroughly monitor these patients during pregnancy for disease activity and embryonic well-being. Providers encounter diagnostic and therapeutic challenges for treating pregnant women with active LN. Steroids, hydroxychloroquine, azathioprine, and calcineurin inhibitors are medications considered safe during pregnancy. Biologic agents have been administered during pregnancy; however, there is little availability of safety data, and MMF and cyclophosphamide should be avoided due to teratogenicity.

Fortunately, pregnancy outcomes appear to be improving for most women with SLE and women with a history of LN, and a successful pregnancy is possible, especially if the disease is in remission. Women with LN should be counseled on the risks and benefits of pregnancy, the teratogenicity of medications, and contraception to ensure that pregnancies are planned and have the highest probability of success.

References

[1] Dalal DS, Patel KA, Patel MA. Systemic lupus erythematosus and pregnancy: a brief review. J Obstet Gynaecol India 2019;69(2):104−9.

[2] Kattah AG, Garovic VD. Pregnancy and lupus nephritis. Semin Nephrol 2015;35(5):487−99.

[3] Hansen RB, Falasinnu T, Faurschou M, Jacobsen S, Simard JF. Risk of end-stage renal disease in patients with systemic lupus erythematosus and diabetes mellitus: a Danish nationwide cohort study. Arthritis Care Res 2023;75(9):1871−7.

[4] Zen M, Ghirardello A, Iaccarino L, Tonon M, Campana C, Arienti S, et al. Hormones, immune response, and pregnancy in healthy women and SLE patients. Swiss Med Wkly 2010;140(13−14): 187−201.

[5] Smyth A, Oliveira GH, Lahr BD, Bailey KR, Norby SM, Garovic VD. A systematic review and meta-analysis of pregnancy outcomes in patients with systemic lupus erythematosus and lupus nephritis. Clin J Am Soc Nephrol 2010;5(11):2060−8.

[6] Ruiz-Irastorza G, Lima F, Alves J, Khamashta MA, Simpson J, Hughes GR, et al. Increased rate of lupus flare during pregnancy and the puerperium: a prospective study of 78 pregnancies. Br J Rheumatol 1996;35(2):133−8.

[7] Hladunewich MA. Chronic kidney disease and pregnancy. Semin Nephrol 2017;37(4):337−46.

[8] Bramham K, Lightstone L. Pre-pregnancy counseling for women with chronic kidney disease. J Nephrol 2012;25(4):450−9.

[9] Imbasciati E, Tincani A, Gregorini G, Doria A, Moroni G, Cabiddu G, et al. Pregnancy in women with preexisting lupus nephritis: predictors of fetal and maternal outcome. Nephrol Dial Transplant 2009;24(2):519−25.

[10] Bramham K, Hunt BJ, Bewley S, Germain S, Calatayud I, Khamashta MA, et al. Pregnancy outcomes in systemic lupus erythematosus with and without previous nephritis. J Rheumatol 2011;38(9): 1906−13.

[11] Bertsias GK, Tektonidou M, Amoura Z, Aringer M, Bajema I, Berden JH, et al. Joint European League Against Rheumatism and European Renal Association-European Dialysis and Transplant Association (EULAR/ERA-EDTA) recommendations for the management of adult and paediatric lupus nephritis. Ann Rheum Dis 2012;71(11):1771−82.

[12] Gianfreda D, Quaglini S, Frontini G, Raffiotta F, Messa P, Moroni G. Does pregnancy have any impact on long term damage accrual and on the outcome of lupus nephritis? J Autoimmun 2017; 84:46−54.

[13] Orefice R. Immunology and the immunological response in pregnancy. Best Pract Res Clin Obstet Gynaecol 2021;76:3−12.

[14] Abu-Raya B, Michalski C, Sadarangani M, Lavoie PM. Maternal immunological adaptation during normal pregnancy. Front Immunol 2020;11:575197.

[15] Zhang S, Han X, Liu W, Wen Q, Wang J. Pregnancy in patients with systemic lupus erythematosus: a systematic review. Arch Gynecol Obstet 2023;308(1):63−71.

[16] Braga A, Neves E, Guimarães J, Braga J, Vasconcelos C. The dynamics of Th17/Treg ratio in SLE patients during pregnancy. J Reprod Immunol 2022;151:103622.

[17] Tower C, Crocker I, Chirico D, Baker P, Bruce I. SLE and pregnancy: the potential role for regulatory T cells. Nat Rev Rheumatol 2011;7(2):124−8.

[18] Gluhovschi C, Gluhovschi G, Petrica L, Velciov S, Gluhovschi A. Pregnancy associated with systemic lupus erythematosus: immune tolerance in pregnancy and its deficiency in systemic lupus erythematosus − an immunological dilemma. J Immunol Res 2015;2015:241547.

[19] Zahiri L, Malekmakan L, Masjedi F, Habibagahi Z, Habibagahi M. Association between IL-17A, FOXP3, and CTLA4 genes expression and severity of lupus nephritis. Iran J Kidney Dis 2022;1(1): 13−23.

[20] Varghese S, Crocker I, Bruce IN, Tower C. Systemic lupus erythematosus, regulatory T cells and pregnancy. Expert Rev Clin Immunol. 2011;7(5):635−48.

[21] Fu B, Tian Z, Wei H. TH17 cells in human recurrent pregnancy loss and pre-eclampsia. Cell Mol Immunol 2014;11(6):564−70.

[22] Maynard S, Guerrier G, Duffy M. Pregnancy in women with systemic lupus and lupus nephritis. Adv Chronic Kidney Dis 2019;26(5):330–7.
[23] Pastore DEA, Costa ML, Parpinelli MA, Surita FG. A critical review on obstetric follow-up of women affected by systemic lupus erythematosus. Rev Bras Ginecol Obstet 2018;40(4):209–24.
[24] Knight CL, Nelson-Piercy C. Management of systemic lupus erythematosus during pregnancy: challenges and solutions. Open Access Rheumatol 2017;9:37–53.
[25] Liu Y, Zhang Y, Wei Y, Yang H. Effect of hydroxychloroquine on preeclampsia in lupus pregnancies: a propensity score-matched analysis and meta-analysis. Arch Gynecol Obstet 2021;303(2):435–41.
[26] Mohammed RHA, Mumtaz H, Sangah AB, Shaikh SS, Nasir N, Jabeen S. Pregnancy in lupus: an updated consensus to guide best practice strategies. Egypt Rheumatol Rehabil 2022;49(1):67.
[27] Shakya S, Yang Z. New-onset systemic lupus erythematosus during pregnancy: a challenge in diagnosis. J Reproduct Health Med 2016;2(2):63–6.
[28] Sammaritano LR, Bermas BL, Chakravarty EE, Chambers C, Clowse MEB, Lockshin MD, et al. 2020 American College of Rheumatology Guideline for the Management of Reproductive Health in Rheumatic and Musculoskeletal Diseases. Arthritis Care Res 2020;72(4):461–88.
[29] Clowse ME, Jamison M, Myers E, James AH. A national study of the complications of lupus in pregnancy. Am J Obstet Gynecol 2008;199(2):127.e1–6.
[30] Clowse ME. Lupus activity in pregnancy. Rheum Dis Clin N Am. 2007;33(2):237–52.
[31] Saavedra MA, Cruz-Reyes C, Vera-Lastra O, Romero GT, Cruz-Cruz P, Arias-Flores R, et al. Impact of previous lupus nephritis on maternal and fetal outcomes during pregnancy. Clin Rheumatol 2012;31(5):813–9.
[32] Chakravarty EF, Colón I, Langen ES, Nix DA, El-Sayed YY, Genovese MC, et al. Factors that predict prematurity and preeclampsia in pregnancies that are complicated by systemic lupus erythematosus. Am J Obstet Gynecol 2005;192(6):1897–904.
[33] Gladman DD, Tandon A, Ibañez D, Urowitz MB. The effect of lupus nephritis on pregnancy outcome and fetal and maternal complications. J Rheumatol 2010;37(4):754–8.
[34] Dao KH, Bermas BL. Systemic lupus erythematosus management in pregnancy. Int J Womens Health 2022;14:199–211.
[35] Nili F, McLeod L, O'Connell C, Sutton E, McMillan D. Maternal and neonatal outcomes in pregnancies complicated by systemic lupus erythematosus: a population-based study. J Obstet Gynaecol Can 2013;35(4):323–8.
[36] Antovic A, Sennström M, Bremme K, Svenungsson E. Obstetric antiphospholipid syndrome. Lupus Sci Med 2018;5(1):e000197.
[37] Zamora-Medina MC, Orozco-Guillén OA, Domínguez-Quintana M, Romero-Diaz J. Systemic lupus erythematosus and pregnancy: strategies before, during and after pregnancy to improve outcomes. Rev Colomb Reumatol 2021;28:53–65.
[38] Sammaritano LR. Management of systemic lupus erythematosus during pregnancy. Annu Rev Med 2017;68(1):271–85.
[39] Chen X, Di W, Ye L, Hu Y, Jiang M, Wu J, et al. Association of preterm outcome with maternal systemic lupus erythematosus: a retrospective cohort study. Ital J Pediatr 2023;49(1):43.
[40] Radu A, Dudu SC, Ciobanu A, Peltecu G, Iancu G, Botezatu R, et al. Pregnancy management in women with antiphospholidic syndrome. Maedica (Bucur). 2019;14(2):148–60.
[41] Attia DH, Mokbel A, Haggag HM, Naeem N. Pregnancy outcome in women with active and inactive lupus nephritis: a prospective cohort study. Lupus 2019;28(7):806–17.
[42] Wagner SJ, Craici I, Reed D, Norby S, Bailey K, Wiste HJ, et al. Maternal and foetal outcomes in pregnant patients with active lupus nephritis. Lupus 2009;18(4):342–7.
[43] Chen Y, Li K, Zhang H, Liu Z, Chen D, Yang L, et al. Good pregnancy outcomes in lupus nephritis patients with complete renal remission. Nephrol Dial Transplant 2022;37(10):1888–94.
[44] Clowse ME, Magder LS, Witter F, Petri M. Early risk factors for pregnancy loss in lupus. Obstet Gynecol 2006;107(2 Pt 1):293–9.
[45] Carmona F, Font J, Moga I, Làzaro I, Cervera R, Pac V, et al. Class III–IV proliferative lupus nephritis and pregnancy: a study of 42 cases. Am J Reprod Immunol 2005;53(4):182–8.

[46] Koh JH, Ko HS, Lee J, Jung SM, Kwok SK, Ju JH, et al. Pregnancy and patients with preexisting lupus nephritis: 15 years of experience at a single center in Korea. Lupus 2015;24(7):764–72.
[47] Clark CA, Spitzer KA, Laskin CA. Decrease in pregnancy loss rates in patients with systemic lupus erythematosus over a 40-year period. J Rheumatol 2005;32(9):1709–12.
[48] Wu J, Ma J, Zhang WH, Di W. Management and outcomes of pregnancy with or without lupus nephritis: a systematic review and meta-analysis. Ther Clin Risk Manag 2018;14:885–901.
[49] Shen M, Tan H, Zhou S, Smith GN, Walker MC, Wen SW. Trajectory of blood pressure change during pregnancy and the role of pre-gravid blood pressure: a functional data analysis approach. Sci Rep 2017;7(1):6227.
[50] Cooper WO, Hernandez-Diaz S, Arbogast PG, Dudley JA, Dyer S, Gideon PS, et al. Major congenital malformations after first-trimester exposure to ACE inhibitors. N Engl J Med 2006;354(23): 2443–51.
[51] Flint J, Panchal S, Hurrell A, van de Venne M, Gayed M, Schreiber K, et al. BSR and BHPR guideline on prescribing drugs in pregnancy and breastfeeding-Part II: analgesics and other drugs used in rheumatology practice. Rheumatology 2016;55(9):1698–702.
[52] Kattah AG, Garovic VD. The management of hypertension in pregnancy. Adv Chronic Kidney Dis 2013;20(3):229–39.
[53] Morales E, Galindo M, Trujillo H, Praga M. Update on lupus nephritis: looking for a new vision. Nephron 2021;145(1):1–13.
[54] Mariotti V, Marconi AM, Pardi G. Undesired effects of steroids during pregnancy. J Matern Fetal Neonatal Med 2004;16(Suppl. 2):5–7.
[55] Webster P, Wardle A, Bramham K, Webster L, Nelson-Piercy C, Lightstone L. Tacrolimus is an effective treatment for lupus nephritis in pregnancy. Lupus 2014;23(11):1192–6.
[56] Michael A, Radwan AA, Ali AK, Abd-Elkariem AY, Shazly SA. Use of antenatal fluorinated corticosteroids in management of congenital heart block: systematic review and meta-analysis. Eur J Obstet Gynecol Reprod Biol X 2019;4:100072.
[57] Abarientos C, Sperber K, Shapiro DL, Aronow WS, Chao CP, Ash JY. Hydroxychloroquine in systemic lupus erythematosus and rheumatoid arthritis and its safety in pregnancy. Expert Opin Drug Saf 2011;10(5):705–14.
[58] Clowse MEB, Eudy AM, Balevic S, Sanders-Schmidler G, Kosinski A, Fischer-Betz R, et al. Hydroxychloroquine in the pregnancies of women with lupus: a meta-analysis of individual participant data. Lupus Sci Med 2022;9(1).
[59] Leroux M, Desveaux C, Parcevaux M, Julliac B, Gouyon JB, Dallay D, et al. Impact of hydroxychloroquine on preterm delivery and intrauterine growth restriction in pregnant women with systemic lupus erythematosus: a descriptive cohort study. Lupus 2015;24(13):1384–91.
[60] Seo MR, Chae J, Kim YM, Cha HS, Choi SJ, Oh S, et al. Hydroxychloroquine treatment during pregnancy in lupus patients is associated with lower risk of preeclampsia. Lupus 2019;28(6):722–30.
[61] Do SC, Rizk NM, Druzin ML, Simard JF. Does hydroxychloroquine protect against preeclampsia and preterm delivery in systemic lupus erythematosus pregnancies? Am J Perinatol 2020;37(9):873–80.
[62] Clowse ME, Magder L, Witter F, Petri M. Hydroxychloroquine in lupus pregnancy. Arthritis Rheum 2006;54(11):3640–7.
[63] Clowse MEB, Eudy AM, Balevic S, Sanders-Schmidler G, Kosinski A, Fischer-Betz R, et al. Hydroxychloroquine in the pregnancies of women with lupus: a meta-analysis of individual participant data. Lupus Sci Med. 2022;9(1):e000651.
[64] Natekar A, Pupco A, Bozzo P, Koren G. Safety of azathioprine use during pregnancy. Can Fam Physician 2011;57(12):1401–2.
[65] Cleary BJ, Källén B. Early pregnancy azathioprine use and pregnancy outcomes. Birth Defects Res A Clin Mol Teratol 2009;85(7):647–54.
[66] Götestam Skorpen C, Hoeltzenbein M, Tincani A, Fischer-Betz R, Elefant E, Chambers C, et al. The EULAR points to consider for use of antirheumatic drugs before pregnancy, and during pregnancy and lactation. Ann Rheum Dis 2016;75(5):795–810.

[67] Sheikholeslami M, Hajialilo M, Rasi Hashemi SS, Malek Mahdavi A, Gojazadeh M, Khabbazi A. Low dose cyclosporine A in the treatment of resistant proliferative lupus nephritis. Mod Rheumatol 2018;28(3):523–9.
[68] Hannah J, Casian A, D'Cruz D. Tacrolimus use in lupus nephritis: a systematic review and meta-analysis. Autoimmun Rev 2016;15(1):93–101.
[69] Alsuwaida A. Successful management of systemic lupus erythematosus nephritis flare-up during pregnancy with tacrolimus. Mod Rheumatol 2011;21(1):73–5.
[70] Suzuki K, Uno S, Wakasugi N. Tacrolimus use and renal function in pregnancy with lupus nephritis: analysis of post-marketing surveillance data in Japan. Mod Rheumatol 2023;33(5):944–52.
[71] Clowse ME, Magder L, Petri M. Cyclophosphamide for lupus during pregnancy. Lupus 2005;14(8):593–7.
[72] Hahn KM, Johnson PH, Gordon N, Kuerer H, Middleton L, Ramirez M, et al. Treatment of pregnant breast cancer patients and outcomes of children exposed to chemotherapy in utero. Cancer 2006;107(6):1219–26.
[73] Gupta R, Deepanjali S, Thabah MM, Kaur P, Gupta S. Successful twin pregnancy while on cyclophosphamide therapy in a patient with lupus nephritis. Rheumatol Int 2009;29(12):1503–5.
[74] Férnandez M, Andrade R, Alarcón GS. Cyclophosphamide use and pregnancy in lupus. Lupus 2006;15(1):59.
[75] Diav-Citrin O, Otcheretianski-Volodarsky A, Shechtman S, Ornoy A. Pregnancy outcome following gestational exposure to TNF-alpha-inhibitors: a prospective, comparative, observational study. Reprod Toxicol 2014;43:78–84.
[76] Rovin BH, Furie R, Latinis K, Looney RJ, Fervenza FC, Sanchez-Guerrero J, et al. Efficacy and safety of rituximab in patients with active proliferative lupus nephritis: the Lupus Nephritis Assessment with Rituximab study. Arthritis Rheum 2012;64(4):1215–26.
[77] Martínez López JA, García Vivar ML, Cáliz R, Freire M, Galindo M, Hernández MV, et al. Recommendations for the evaluation and management of patients with rheumatic autoimmune and inflammatory diseases during the reproductive age, pregnancy, postpartum and breastfeeding. Reumatol Clin 2017;13(5):264–81.
[78] Sangle SR, Lutalo PM, Davies RJ, Khamashta MA, D'Cruz DP. B-cell depletion therapy and pregnancy outcome in severe, refractory systemic autoimmune diseases. J Autoimmun 2013;43:55–9.
[79] Allen ME, Rus V, Szeto GL. Leveraging heterogeneity in systemic lupus erythematosus for new therapies. Trends Mol Med 2021;27(2):152–71.
[80] Ginzler EM, Wallace DJ, Merrill JT, Furie RA, Stohl W, Chatham WW, et al. Disease control and safety of belimumab plus standard therapy over 7 years in patients with systemic lupus erythematosus. J Rheumatol 2014;41(2):300–9.
[81] Danve A, Perry L, Deodhar A. Use of belimumab throughout pregnancy to treat active systemic lupus erythematosus: a case report. Semin Arthritis Rheum 2014;44(2):195–7.
[82] Monova D, Belovezhdov N, Altunkova I, Monov S. Intravenous immunoglobulin G in the treatment of patients with chronic glomerulonephritis: clinical experience lasting 15 years. Nephron 2002;90(3):262–6.
[83] Micheloud D, Nuño L, Rodríguez-Mahou M, Sánchez-Ramón S, Ortega MC, Aguarón A, et al. Efficacy and safety of Etanercept, high-dose intravenous gammaglobulin and plasmapheresis combined therapy for lupus diffuse proliferative nephritis complicating pregnancy. Lupus 2006;15(12):881–5.

CHAPTER 11

Pregnancy with preexisting kidney disease

Juan Ignacio Silesky-Jimenez
San Juan de Dios Hospital, Caja Costarricense del Seguro Social and CIMA Hospital, San José, Costa Rica

Introduction

Chronic kidney disease is a condition that is associated with several negative outcomes during pregnancy [1,2], which can occur in about 6% of women of reproductive age and up to 3% of pregnant women [3].

Chronic kidney disease is defined as abnormalities in kidney structure or function, lasting more than 3 months, with health implications, which is determined by the decrease in the glomerular filtration rate (GFR) which is also associated with other abnormalities such as proteinuria, tubular disorders, and structural/congenital abnormalities [4].

Women with advanced kidney disease have poor fertility; however, cases of pregnant women with endstage kidney disease (ESKD) have been described [5].

There are multiple causes of kidney disease in this group of patients, such as: congenital kidney and urinary tract abnormalities, hypertensive nephrosclerosis, diabetic nephropathy and glomerulopathies such as IgA nephropathy and lupus nephritis.

The adverse outcomes that the population of pregnant women present are: preeclampsia, premature births, children small for gestational age (SCA), low birth weight (LBW), Caesarean sections (C-section), and anemia [6—14].

Pregnant patients with mild chronic kidney disease (CKD 1—2) usually have a favorable postpartum renal prognosis. Patient with severe renal dysfunction (CKD3b-5) at conception, usually associate a significant decrease in renal function during pregnancy is observed, in patients with severe renal dysfunction (CKD 3b-5) at conception [16—18].

Some cases require intensive dialysis to improve the maternal—fetal prognosis, in developed countries where there is a great supply of such support [14,15]; and in many cases, they remain dependent on it [11,14].

Physiological changes in pregnancy

In normal women, physiological and immunological changes occur during pregnancy, which help the mother tolerate this process.

The Kidney of the Critically Ill Pregnant Woman
ISBN 978-0-443-21473-8,
https://doi.org/10.1016/B978-0-443-21473-8.00006-9

As is known, the hemodynamic changes that occur are an increase in blood volume, decrease in vascular resistance, and increase in cardiac output. The above is associated with an increase in vasodilators such as nitric oxide and relaxin, as well as a relative resistance to vasoconstrictors such as angiotensin II. These changes produce a decrease in blood pressure, which is maximum in week 20.

Glomerular filtration experiences an increase of 50%; this hyperfiltration manifests itself with a decrease in creatinine levels, typically being between 0.4 and 0.6 mg/dL, which, together with changes in glomerular permeability, produces an increase in proteinuria that it can reach 250 mg/dL, its limit being 300 mg/dL. Furthermore, due to the effects of progesterone on the relaxation of smooth muscle and the mechanical effect of the gravid uterus, the presence of physiological hydronephrosis and retention of urine in the collecting system is common [19].

From an immunological point of view, there are changes in innate and adaptive immunity, in order to adapt to fetal antigens, trophoblastic invasion, and the formation of the placenta. These changes may be important, in the context of patients with autoimmune diseases. In normal pregnancy, there is a change in phenotype from Th1 (cell-mediated immunity) to Th2 (humoral-mediated immunity). Another change is the increase in regulatory T cells (T regs) that promote immune tolerance. Quantitative and quantitative alterations in T regs may correlate with an increase in pregnancy complications in patients with systemic lupus erythematosus (SLE), such as preeclampsia and poor fetal and maternal prognoses [20].

In summary, the physiologic changes in pregnancy are:

1. Increase: Blood volume
   Cardiac output
   Levels of nitric oxide and relaxin
   Relative resistance to vasoconstrictors
   GRF by 50%
   Urine protein losses
   TH2 phenotype
   Circulation of T regs
2. Decrease: Systemic blood pressure
   Serum creatinine
   Systemic vascular resistance

Pregnancy is an intense, real-time stress on the kidney, where the important physiological changes described previously can unmask or worsen underlying kidney conditions, and where many young women have their kidney function determined for the first time [21]. Kidney disease in pregnancy involves several heterogeneous conditions, often unrecognized or unreported, that have an adverse impact on the maternal—fetal prognosis [22]. Deterioration of renal function is an independent factor of poor obstetric prognosis, particularly in preterm delivery and growth restriction [8,23—25].

A chronic kidney disease in pregnancy

Chronic kidney disease in the context of pregnancy has an impact on different topics:

1. It increases the risk of negative maternal–fetal effects such as an increase in the prevalence of preeclampsia, preterm births, LBW, and an increase in mortality.
2. Deterioration of renal function in patients with advanced chronic kidney disease, especially those with antepartum creatinine greater than 2.0 mg/dL [26].
3. It may produce an increase in albuminuria, despite having kidney function close to normal [27].

In this group of patients, an evaluation by high-risk obstetrics and nephrology before pregnancy is essential, in order to determine its risk for the patient, since it is known that:

1. Women with mild chronic kidney disease with serum creatinine <1.4 mg/dL may have a good maternal–fetal prognosis.
2. Women with advanced chronic kidney disease with serum creatinine between 1.4 and 2.9 mg/dL have a high risk of maternal–fetal complications.
3. Women with serum creatinine ≥ 3.0 mg/dL may permanently lose their kidney function with pregnancy.

In addition, the specific cause or pathology of kidney damage presents specific risks as in the case of lupus nephritis, diabetes mellitus (DM), etc. [28].

Pregnancy in end-stage renal disease in pregnancy

It is well known that patients on dialysis have decreased fertility due to irregular or absent menstrual cycles. Despite the above, in this context, there is the possibility of pregnancy, especially in patients with intensified dialysis programs [29].

For the diagnosis of pregnancy, serum hCG can be used. This determination may be slightly elevated in these patients; however, its increase every 48–72 hours may indicate the presence of pregnancy, which can be confirmed by ultrasound.

For pregnant patients, intensive renal support must be achieved to bring urea nitrogen close to normal, ideally with goals of 36 hours per week and seric ureic nitrogen less than 50 mg/dL. In this group of patients, maternal mortality is low and an improvement in results has been observed when the frequency and duration of renal support sessions is increased [30].

Pregnant patients for transplant and donation candidates

In patients with chronic kidney failure eligible for kidney transplant, it is recommended to wait to become pregnant until a successful transplant is achieved, because after achieving the transplant, fertility improves and the risk of complications such as preeclampsia and preterm births are reduced.

It is important to remember that pregnancy is a sensitizing event, which limits the possibility of finding compatible donors because anti-HLA antibodies are formed. The

KDIGO (**K**idney **D**isease: **I**mproving **G**lobal **O**utcomes) [31] recommendation is that women wait at least 1 year posttransplant before becoming pregnant, when kidney function is stable. However, it is recommended to wait 2 years posttransplant to reduce the risk of graft failure.

Regarding the evolution of patients with kidney transplants and who have achieved pregnancy, there is a metaanalysis published in 2011, which included 50 studies, with a total of 4700 pregnancies in 3570 patients with kidney transplants [32]. It is documented that the number of live births is similar to that of the general population (73.5% vs. 66.7%), but there is a significant increase in complications such as preeclampsia (27% vs. 3.8%), gestational diabetes (8% vs. 3.9%), and preterm births (45.6% vs. 12.5%). In other cohort studies, the risk of graft loss is low (5.8% at one year and 6.9% at 5 years).

During pregnancy, these patients require very close monitoring by obstetricians and nephrologists. Immunosuppressive treatment must not have contraindications during pregnancy, such as the use of azathioprine, tacrolimus, cyclosporine, and prednisone; or combinations of these. Special consideration should be given to tacrolimus, since during pregnancy toxicity may occur despite having adequate blood levels, since there is no correlation with the free levels of this drug [33].

Kidney donor patients must have special surveillance because they have a higher risk of hypertensive complications, such as preeclampsia and pregnancy-induced hypertension. However, there was no increase in preterm births or LBW [34].

Specific disease to produce chronic kidney disease in pregnancy

Diabetic nephropathy

Diabetic nephropathy is a condition that is characterized by being slow and gradual, manifested by the development of high blood pressure, albuminuria, and loss of GFR. Its incidence has been estimated at 6% of pregnant women with Type I DM, since due to its gradual course, it is rare in women with Type II DM of reproductive age [35].

The risk of complications will depend on the degree of glomerular involvement, evidenced by the patient's degree of kidney function before starting pregnancy. Women with serum creatinine >1.4 mg/dL have a high risk of kidney deterioration and progression to ESRD with pregnancy [36].

ACE inhibitors and ARBs cannot be used in pregnancy, but their use before pregnancy, at least 3–6 months, has a protective effect [37]. However, without these medications and with the effects of pregnancy on kidney function mentioned above, they produce increases in albuminuria up to seven times, even exceeding 3 g/d. This condition can make the diagnosis of preeclampsia difficult. In any case, this group of women with the characteristics of being pregnant with diabetic nephropathy may have a risk of preeclampsia up to 2/3 if it occurs. Therefore, although there is no evidence, this group of patients can benefit from the use of aspirin as a prophylactic measure, which is also low risk of harm.

All women return to their baseline state of kidney function at 12 weeks postpartum.

From the point of view of pregnancy, diabetic nephropathy presents an increase in complications such as spontaneous abortions, congenital malformations, preterm births, macrosomia, and perinatal mortality. Due to this, there must be a prior multidisciplinary assessment (high-risk obstetrics, nephrology, and endocrinology), in which the risk is assessed, glycemic control is optimized, and all precautions are taken to achieve a successful pregnancy.

The American Diabetes Association recommends a glycosylated Hb $< 6.5\%$ at least 6 months before getting pregnant, in order to improve the prognosis. The basis for achieving these objectives is with the use of insulin; however, in type II DM, oral hypoglycemic agents can be used when successful glycemic control is obtained. Furthermore, patients with associated diabetic retinopathy, especially if it is moderate or severe, must be closely monitored because it can worsen during pregnancy; especially those patients with high glycosylated Hb levels whose glycemic values are quickly controlled at the beginning of pregnancy [38].

Lupus nephritis

Women diagnosed with SLE of reproductive age require an early evaluation when the diagnosis is made, to determine the maternal—fetal risk in the event of a desired pregnancy. Ideally, SLE should be inactive for at least 6 months before pregnancy, in order to have the best maternal—fetal outcomes.

The activity of the disease must be measured with the available tools, such as biological markers, kidney function, and the degree of proteinuria; in order to have a timely diagnosis of disease outbreaks, as well as monitor any complications during pregnancy, in order to categorize high-risk patients and thus be able to reduce the progression of the disease and improve results.

It is striking that extrarenal outbreaks occur between the second and third trimester, with renal outbreaks being more frequent in the postpartum period. The incidence of severe outbreaks during pregnancy occurs between 3% and 5% of cases.

There are some predictors of possible SLE outbreaks in the kidney such as:

1. Immunological activity during conception, especially with the determination of low levels of C3 and anti-DNA antibodies.
2. The presence of low levels of C4 and elevations of anti-C1 are associated with early outbreaks, during the second and third trimester.
3. Being overweight increases the risk of flare-ups due to SLE, especially in the third trimester and postpartum [39,40].

The presence of a kidney flare is not an absolute indication to terminate the pregnancy.

Regarding maternal and fetal complications in SLE with lupus nephropathy, it has been observed that:

1. There is an increase in flares due to SLE, hypertension and preeclampsia.
2. In patients with known lupus nephropathy, they have a higher risk of preterm delivery, and an early presentation of preeclampsia.
3. They have an estimated mortality of 1%.
4. According to the PROMISSE study (Predictors of Pregnancy Outcome: Biomarkers in Antiphospholipid Antibody Syndrome and Systemic Lupus Erythematosus), with 385 pregnant women with SLE, 19% presented poor outcomes during pregnancy such as: 4% fetal deaths, 1% neonatal death, 9% preterm birth, and SGA by 10%. The presence of an anticoagulant lupic, an activity score >1, the use of antihypertensives, and counts platelet <100,000/μL are independent predictors of adverse outcomes [39].
5. SLE activity in the last year before pregnancy is a good predictor of fetal outcomes, such that: (a) Abortions are associated with the amount of steroids used in the year before pregnancy. (b) The number of SLE outbreaks in the year before pregnancy are direct predictors of fetal outcomes (increase miscarriages, stillbirths and preterm births) (c) The presence of antiphospholipid syndrome and antidouble-chain antibodies DNA before pregnancy is associated with preterm births. d. The presence of neonatal SLE, manifested as heart block complete congenital, has an incidence of 1%–2% after exposure to the SSA/Ro and SSB/La antibodies. This incidence increases to 20%, if there is a maternal history of previous births with children with neonatal SLE. It is important to perform anti-Ro (SSA) and anti-La (SSB) determinations in lupus pregnant women.
6. Antiphospholipid antibodies are positive in 25% of cases. Pregnant women with SLE, which is also associated with fetal loss and a relative increase in preeclampsia. The determination of these antibodies should be done at the beginning of pregnancy. Management is based on anticoagulation by low molecular weight heparins; with an increase in complications thromboembolic events due to a state of hypercoagulability, an effect mechanical obstructive effect of the pregnant uterus on venous return and an increase in proteinuria in patients with glomerular disease, which can reach ranges of nephrotic syndrome, which is also associated with a risk increased thromboembolic phenomena. Anticoagulation is especially indicated in women with SLE with antiphospholipid antibodies with a history of thrombotic events or those who have a history of three or more pregnancy loss or death [41].

As for the medications usually used in this condition, some medications are safe, but it is important to remember that:

1. Hydroxychloroquine should be continued during pregnancy, in order to maintain inactivity of lupus nephritis and the risk of systemic flares. Its discontinuation increases the risk of starting lupus activity and the appearance of outbreaks of the disease, which can require high doses of steroids. It has no risk of teratogenicity.

2. Glucocorticoids: these medications increase the risk of gestational, lip and palate diabetes cleft, as well as premature rupture of membranes.
3. Azathioprine: does not increase the risk of teratogenicity.
4. Cyclosporine: increases the risk of cholestasis.
5. Tacrolimus: increases the risk of high blood pressure and gestational diabetes.

Other medications are dangerous such as:

1. Cyclophosphamide: produces fetal malformations and increases the frequency of pregnancy loss.
2. Mycophenolate mofetil: it is a teratogenic drug (abnormalities of the lip, palate, and hearing), with an increase in the frequency of pregnancy loss. Rituximab may transiently cause B-cell depletion in the fetus [42].

Nephrotic syndrome in pregnancy

This syndrome can be difficult to diagnose during pregnancy and may require taking a kidney biopsy to determine if it is a kidney pathology itself, or some pathology associated with pregnancy such as preeclampsia, especially if its appearance is before pregnancy week 20.

Among the possible causes, we can list:

1. **Focal segmental glomerulosclerosis**: This histological pattern has primary and secondary causes. Primary focal segmental glomerulosclerosis is characterized by diffuse foot process effacement (>80%), secondary FSGS presents moderate foot process effacement, primary FSGS producing an acute or subacute nephrotic syndrome, while secondary FSGS presents decreased renal function, with proteinuria in non-nephrotic ranges. These patients, with primary FSGS, benefit from the use of glucocorticoids and/or calcineurin inhibitors [43].
2. **Minimal change disease:** It is a cause of nephrotic syndrome, but it is rare during pregnancy. The management of this pathology may require the use of antihypertensives and glucocorticoids. Diuretics should be used with care, since theoretically, they can cause intravascular depletion and/or cause systemic vasoconstriction and cause placental hypoperfusion in preeclampsia; some recommend sodium restriction and fluid restriction in nephrotic syndrome. Patients who, before pregnancy, are known carriers of FSGS and minimal change disease must be in remission before becoming pregnant. Patients with relapses can be treated with azathioprine or calcineurin inhibitors.
3. **Membranous nephropathy** in women of reproductive age, it is secondary to other pathologies such as SLE, Hepatitis B, Hepatitis C, syphilis, drugs (biological agents and NSAIDs), or less commonly due to tumors, which could be primary kidney tumors. Many of the medications used in this disease are contraindicated in pregnancy, such as ACE inhibitors, lipid-lowering therapy, warfarin, and cyclophosphamide [44].

IgA nephropathy

This type of nephropathy manifests itself between the second and third decade of life, so it affects women of reproductive age.

Women with IgA nephropathy do not experience long-term impacts on kidney function with pregnancy, in women with mild disease (CKD stage 1 and 2); however, they do show an increase in pregnancy complications such as pregnancy loss, preterm births, LBW, preeclampsia/eclampsia. It is important to clarify that women with this type of nephropathy in association with an eGFR <60 mL/minutes/1.73 m^2 or >1 g/d of proteinuria may have a higher risk of progression of kidney disease [45,46].

Patients with nephropathy with mild, stable, or slowly progressive IgA do not need immunosuppressive medications; ACEi can be suspended before pregnancy and if they require any immunosuppressive treatment, they can be changed to safe options for the fetus. Steroids are allowed during pregnancy, if necessary to deepen immunosuppression in active disease [47].

Alport syndrome and pregnancy

Alport syndrome is a rare genetic disease in which there is an abnormality in the genes that encode the collagen IV chains, which are a3, a4, a5 (COL4A3, COL4A4, and COL4A5) [48] that produce a defect in the structure and function of the basement membrane in different organs, such as the glomerulus. The above manifests itself from a clinical point of view in microscopic hematuria and can evolve to proteinuria and end-stage renal disease (ESRD), associated with neurosensory deafness and ocular defects [49,50].

This strange disease has been reported in a few series of patients, in which the following has been observed:

1. Patients with isolated microscopic hematuria have an uncomplicated pregnancy.
2. Patients with mild proteinuria at the time of conception present an increase in it in the last trimester, even to nephrotic ranges, associated with fluid overload that warranted hospitalization and early delivery. Most neonates presented LBW, preeclampsia, and persistent deterioration in renal function.
3. It has been described that patient with this pathology, with chronic kidney damage, can develop ESRD; so, advice is important.
4. Some patients with normal blood pressure and kidney function improved proteinuria after termination of pregnancy [51].

These patients must be evaluated and receive advice before starting pregnancy, as well as very close medical follow-up during pregnancy.

High blood pressure and pregnancy

This condition can occur preexisting pregnancy; as a hypertensive syndrome associated with pregnancy (for example preeclampsia), which is common in patients with known kidney disease prior to pregnancy, or the latter on previous chronic high blood pressure, causing an increase in cardiovascular and cerebrovascular complications of the patients, and in maternal–fetal complications. It is also important to keep in mind that patients with these hypertensive diseases have an increased long-term risk of cardiovascular, kidney, and metabolic diseases.

Chronic arterial hypertension is defined as the presence of BP > 140/90 before pregnancy, or before the 20th week of pregnancy on two occasions and that persists beyond the 12th week postpartum. Ideally, corroborated with ambulatory blood pressure monitoring, to exclude white coat hypertension, which is very common in pregnant women [52].

This disease, with adequate treatment and control, is well tolerated during pregnancy, with little probability of maternal–fetal complications; however, if this disease is uncontrolled, there is a high risk of having a poor obstetric prognosis and complications in the mother's target organs.

Patients with chronic arterial hypertension have a worsening of their condition and prognosis if they present superimposed pregnancy-induced hypertension, which occurs in 17%–25% of known hypertensive women versus 2%–5% of the general population [53]. These pregnancies also present an increase in cesarean sections, preterm births, LBW, need for neonatal intensive care, and perinatal deaths [54].

Due to the above, it is of utmost importance to control blood pressure in the antepartum phase, maintaining adequate blood pressure levels, taking the necessary measures to prevent and early diagnose preeclampsia, such as frequent blood pressure monitoring, blood pressure in their homes, education on symptoms and warning signs of preeclampsia and its complications, as well as assessing the effect of the prescribed medication.

Prepregnancy counseling for women with chronic kidney disease

It is of critical importance in women with chronic kidney diseases to have a complete evaluation of their medical condition, to achieve optimization of medical treatment, being clear about what drugs are necessary to control their underlying kidney pathology; before the decision to schedule a pregnancy, in order to provide adequate prepregnancy counseling in which the future risk to the mother and fetus is determined and discussed.

The emotional toll of a complicated pregnancy on the mother and her family should not be underestimated.

This evaluation allows us to clarify several topics for the patient such as:

1. The risk they assume if they become pregnant, so it helps them make a decision.
2. The severity of their illness is clarified, as well as the effects that high blood pressure, proteinuria, and the potentially teratogenic medications they use may present.
3. The vast majority of patients who go through this process have a favorable opinion of this type of help or experience [55].

References

[1] Jha V, Garcia-Garcia G, Iseki K, Li Z, Naicker S, Plattner B, et al. Chronic kidney disease: global dimension and perspectives. Lancet 2013;382(9888):260–72.

[2] Williams D, Davison J. Chronic kidney disease in pregnancy. BMJ 2008;336(7637):211–5.

[3] Webster P, Lightstone L, McKay DB, Josephson MA. Pregnancy in chronic kidney disease and kidney transplantation. Kidney Int 2017;91(5):1047–56.

[4] Chapter 1: Definition and classification of CKD. Kidney Int Suppl 2013;3:19–62.

[5] Holley J, Schmidt R. Changes in fertility and hormone replacement therapy in kidney disease. Adv. Chronic Kidney Dis. 2013;20:240–5.

[6] Zhang JJ, Ma XX, Hao L, Liu LJ, Lv JC, Zhang H. A systematic review and meta-analysis of outcomes of pregnancy in CKD and CKD outcomes in pregnancy. Clin J Am Soc Nephrol 2015;10(11):1964–78.

[7] Nevis IF, Reitsma A, Dominic A, McDonald S, Thabane L, Akl EA, et al. Pregnancy outcomes in women with chronic kidney disease: a systematic review. Clin J Am Soc Nephrol 2011;6(11): 2587–98.

[8] Piccoli GB, Cabiddu G, Attini R, Vigotti FN, Maxia S, Lepori N, et al. Risk of adverse pregnancy outcomes in women with CKD. J Am Soc Nephrol 2015;26(8):2011–22.

[9] Chopra S, Suri V, Aggarwal N, Rohilla M, Keepanasseril A, Kohli HS. Pregnancy in chronic renal insufficiency: single centre experience from North India. Arch Gynecol Obstet 2009;279(5):691–5.

[10] Vázquez-Rodríguez JG, Rivera-Hernández M. Perinatal complications in patients with chronic renal failure. Ginecol Obstet Mex 2011;79(5):261–8.

[11] Sato JL, De Oliveira L, Kirsztajn GM, Sass N. Chronic kidney disease in pregnancy requiring first-time dialysis. Int J Gynaecol Obstet 2010;111(1):45–8.

[12] Ibarra-Hernández M, Orozco-Guillén OA, de la Alcantar-Vallín ML, Garrido-Roldan R, Jiménez-Alvarado MP, Castro KB, et al. Acute kidney injury in pregnancy and the role of underlying CKD: a point of view from México. J Nephrol 2017;30(6):773–80.

[13] Gómez Vázquez JA, MartínezCalva IE, Mendíola Fernández R, Escalera León V, Cardona M, Noyola H. Pregnancy in end-stage renal disease patients and treatment with peritoneal dialysis: report of two cases. Perit Dial Int 2007;27(3):353–8.

[14] Trevisan G, Ramos JG, Martins-Costa S, Barros EJ. Pregnancy in patients with chronic renal insufficiency at Hospital de Clínicas ofPorto Alegre, Brazil. Ren Fail 2004;26(1):29–34.

[15] Hladunewich MA, Hou S, Odutayo A, Cornelis T, Pierratos A, Goldstein M, et al. Intensive hemodialysis associates with improved pregnancy outcomes: a Canadian and United States cohort comparison. J Am Soc Nephrol 2014;25(5):1103–9.

[16] Blom K, Odutayo A, Bramham K, Hladunewich M. Pregnancy and glomerular disease. Clin J Am Soc Nephrol 2017;12:1862–72.

[17] Feng Z, Minard C, Raghavan R. Pregnancy outcomes in advanced kidney disease. Clin Nephrol 2015;83:272–8.

[18] Buyon J, et al. Predictors of pregnancy outcomes in patients with lupus. Ann Intern Med 2015;163:153.

[19] Odutayo A, Hladunewich M. Obstetric nephrology: renal hemodynamic and metabolic physiology in normal pregnancy. Clin J Am Soc Nephrol 2012;7(12):2073–80.

[20] Aluvihare VR, Kallikourdis M, Betz AG. Regulatory T cells mediate maternal tolerance to the fetus. Nat Immunol 2004;5(3):266–71.

[21] Fitzpatrick A, Mohammadi F, Jesudason S. Managing pregnancy in chronic kidney disease: improving outcomes for mother and baby Int J Womens Health 2016;8:273–85.

[22] Piccoli GB, Attini R, Vasario E, Conijn A, Biolcati M, D'Amico F, et al. Pregnancy and chronic kidney disease: a challenge in all CKD stages. Clin J Am Soc Nephrol 2010;5(5):844–55.

[23] Imbasciati E, Gregorini G, Cabiddu G, Gammaro L, Ambroso G, Del Giudice A, et al. Pregnancy in CKD stages 3 to 5: fetal and maternal outcomes. Am J Kidney Dis 2007;49(6):753–62.

[24] Munkhaugen J, Lydersen S, Romundstad PR, Widerøe TE, Vikse BE, Hallan S. Kidney function and future risk for adverse pregnancy outcomes: a populationbased study from HUNT II, Norway. Nephrol Dial Transplant 2009;24(12):3744–50.

[25] Hladunewich MA, Melamed N, Bramham K. Pregnancy across the spectrum of chronic kidney disease. Kidney Int 2016;89(5):995–1007.
[26] Jones DC, Hayslett JP. Outcome of pregnancy in women with moderate or severe renal insufficiency. N Engl J Med 1996;335(4):226.
[27] Zhang JJ, Ma XX, Hao L, et al. A systematic review and meta-analysis of outcomes of pregnancy in CKD and CKD outcomes in pregnancy. Clin J Am Soc Nephrol 2015;10(11):1964–78.
[28] Nevis IF, Reitsma A, Dominic A, et al. Pregnancy outcomes in women with chronic kidney disease: a systematic review. Clin J Am Soc Nephrol 2011;6(11):2587–98.
[29] Hladunewich MA, Hou S, Odutayo A, et al. Intensive hemodialysis associates with improved pregnancy outcomes: a Canadian and United States cohort comparison. J Am Soc Nephrol 2014;25(5): 1103–9.
[30] Piccoli GB, Minelli F, Versino E, et al. Pregnancy in dialysis patients in the new millennium: a systematic review and meta-regression analysis correlating dialysis schedules and pregnancy outcomes. Nephrol Dial Transplant 2016;31(11):1915–34.
[31] Kidney Disease: Improving Global Outcomes Transplant Work Group. KDIGO clinical practice guideline for the care of kidney transplant recipients. Am J Transplant 2009;9(Suppl. 3):S1–155.
[32] Deshpande NA, et al. Pregnancy outcomes in kidney transplant recipients: a systematic review and meta-analysis. Am J Transplant 2011;11(11):2388–404.
[33] Zheng S, et al. Pharmacokinetics of Tacrolimus during pregnancy. Ther Drug Monit 2012;34(6): 660–70.
[34] Garg AX, et al. Gestational hypertension and preeclampsia in living kidney donors. N Engl J Med 2015;372:124–33.
[35] Reece EA, Leguizamon G, Homko C. Pregnancy performance and outcomes associated with diabetic nephropathy. Am J Perinatol 1998;15(7):413–21.
[36] Mackie AD, Doddridge MC, Gamsu HR, et al. Outcome of pregnancy in patients with insulin-dependent diabetes mellitus and nephropathy with moderate renal impairment. Diabet Med 1996; 13(1):90–6.
[37] Hod M, van Dijk DJ, Karp M, et al. Diabetic nephropathy and pregnancy: the effect of ACE inhibitors prior to pregnancy on fetomaternal outcome. Nephrol Dial Transplant 1995;10(12):2328–33.
[38] Chew EY, Mills JL, Metzger BE, et al. Metabolic control and progression of retinopathy. The diabetes in early pregnancy study. National Institute of Child Health and Human Development Diabetes in Early Pregnancy Study. Diab Care 1995;18(5):631.
[39] Kim MY, Buyon JP, Guerra MM, et al. Angiogenic factor imbalance early in pregnancy predicts adverse outcomes in patients with lupus and antiphospholipid antibodies: results of the PROMISSE study. Am J Obstet Gynecol 2016;214(1):108.
[40] Smyth A, Oliveira GH, Lahr BD, et al. A systematic review and meta-analysis of pregnancy outcomes in patients with systemic lupus erythematosus and lupus nephritis. Clin J Am Soc Nephrol 2010;5(11): 2060–8.
[41] Moroni G, Doria A, Giglio E, et al. Fetal outcome and recommendations of pregnancies in lupus nephritis in the 21st century. A prospective multicenter study. J Autoimmun 2016;74:6–12.
[42] Bates SM, Greer IA, Pabinger I, et al. Venous thromboembolism, thrombophilia, antithrombotic therapy, and pregnancy: American College of Chest Physicians evidence-based clinical practice Guidelines (8th Edition). Chest 2008;133(6 Suppl. l):844S–86S.
[43] Nagai Y, Arai H, Washizawa Y, et al. FSGS-like lesions in pre-eclampsia. Clin Nephrol 1991;36(3): 134–40.
[44] De Castro I, Easterling TR, Bansal N, Jefferson JA. Nephrotic syndrome in pregnancy poses risks with both maternal and fetal complications. Kidney Int 2017;91(6):1464–72.
[45] Limardo M, Imbasciati E, Ravani P, et al. Pregnancy and progression of IgA nephropathy: results of an Italian multicenter study. Am J Kidney Dis 2010;56(3):506–12.
[46] Park S, Yoo KD, Park JS, et al. Pregnancy in women with immunoglobulin A nephropathy: are obstetrical complications associated with renal prognosis? Nephrol Dial Transplant 2018;33(3): 459–65.

[47] Liu Y, Ma X, Zheng J, et al. A systematic review and meta-analysis of kidney and pregnancy outcomes in IgA nephropathy. Am J Nephrol 2016;44(3):187–93.
[48] Savige J, Colville D, Rheault M, Gear S, Lennon R, Lagas S, Finlay M, Flinter F. Alport syndrome in women and girls. Clin J Am Soc Nephrol 2016;11:1713–20.
[49] Cosgrove D, Liu S. Collagen IV diseases: a focus on the glomerular basement membrane in Alport syndrome. Matrix Biol 2017;57–58:45–54.
[50] Hudson BG, Tryggvason K, Sundaramoorthy M, Neilson EG. Alport's syndrome, Goodpasture's syndrome, and type IV collagen. New Engl J Med 2003;348:2543–56.
[51] Brunini F, Zaina B, Gianfreda D, Ossola W, Giani M, Fedele L. Alport syndrome and pregnancy: a case series and literature review. Arch Gynecol Obstet 2018;297(6):1421–31.
[52] Reddy S, Jim B. Hypertension and pregnancy: management and future risks. Adv Chronic Kidney Dis 2019;26(2):137–45.
[53] Sibai BM. Chronic hypertension in pregnancy. Obstetr Gynecol 2002;100(2):369–77.
[54] Bramham K, Parnell B, Nelson-Piercy C, Seed PT, Poston L, Chappell LC. Chronic hypertension and pregnancy outcomes: systematic review and meta-analysis. BMJ 2014;348:g2301.
[55] Willes K, Bramham K, Vais A, Harding KR, Chowdhury P, Taulos CJ, Nelson-Piercy C. Pre-pregnancy conunselling for women withchronic kidney disease: a retrospective analysis of nine years experience. Nephrology 2015;16:28.

CHAPTER 12

Hypertensive disorders of pregnancy

Carlos Sánchez[1] and Juan Carlos Barrientos[2]
[1]Critical Care Medicine, King Salman Hospital, Riyadh, Saudi Arabia; [2]High Dependency Obstetric Unit, San Juan de Dios General Hospital, Ciudad de Guatemala, Guatemala

Definition

Hypertensive disorders are the most common medical complication of pregnancy, affecting 6%–8% of all pregnant women. 70% of cases are due to preeclampsia. Preeclampsia is one of the leading causes of maternal death in the world, and a major cause of perinatal death and prematurity [1].

Hypertension in pregnancy is defined as systolic blood pressure of 140 mm Hg or higher or diastolic blood pressure of 90 mm Hg or more on two occasions at least 4 h apart; or systolic blood pressure of 160 mm Hg or more or diastolic blood pressure of 110 mm Hg or more in a single dose for the need to initiate antihypertensive therapy [2,3].

Hypertensive disorders associated with pregnancy could be grouped into four categories [1–3]:

1. Gestational hypertension
2. Chronic hypertension
3. Preeclampsia/eclampsia
4. Preeclampsia superimposed on chronic hypertension.

Gestational hypertension is defined as an elevation in blood pressure after 20 weeks' gestation without proteinuria or systemic findings.

Chronic hypertension is defined as hypertension that exists either before pregnancy or hypertension that develops during pregnancy and fails to resolve by 12 weeks postpartum.

Eclampsia is defined by seizure activity in the presence of preeclampsia. Although it typically presents as a late manifestation after other systemic signs of disease, it can present without other warning signs.

Chronic hypertension superimposed with preeclampsia can present both diagnostic and treatment challenge (proteinuria, increase in liver enzymes, and thrombocytopenia below 100,000/uL), right upper quadrant pain, pulmonary edema, or renal insufficiency. Development of hypertension superimposed with preeclampsia has a worse prognosis for both mother and fetus than either clinical condition alone. This diagnosis is made when a woman with preexisting hypertension after 20 weeks' gestation presents with a sudden exacerbation in hypertension, development of signs or symptoms consistent with preeclampsia [1–3].

The Kidney of the Critically Ill Pregnant Woman
ISBN 978-0-443-21473-8
https://doi.org/10.1016/B978-0-443-21473-8.00009-4

Definition of Preeclampsia, American College of Obstetricians and Gynecologists (ACOG): Preeclampsia is a pregnancy disorder associated with new-onset hypertension, most often occurring after 20 weeks' gestation and often close to term. Although often accompanied by new-onset proteinuria, hypertension and other signs or symptoms of preeclampsia may occur in some women in the absence of proteinuria. In this setting, it is recommended that women with gestational hypertension in the absence of proteinuria be diagnosed with preeclampsia if they have any of the following severe features: thrombocytopenia (platelet count less than 100,000); impaired liver function, indicated by abnormally high blood concentrations of liver enzymes (up to twice the upper limit of normal concentration); persistent severe pain in the right upper quadrant or epigastric pain and not explained by alternative diagnoses; renal impairment (serum creatinine concentration greater than 1.1 mg/dL or a doubling of serum creatinine concentration in the absence of other kidney disease); pulmonary edema; or new-onset headache that does not respond to paracetamol and is not explained by alternative diagnoses or visual disturbances [3,4].

According to its appearance, it can be classified into early preeclampsia that occurs between 20 weeks' and 34 weeks' gestation and late-onset preeclampsia that develops on, or after, 34 weeks. Early onset preeclampsia increases the risk of fetal death (odds ratio 5.8). Additionally, early onset preeclampsia is associated with a significantly increased odds ratio for perinatal death and severe neonatal morbidity (odds ratio 16.4). Although the prognosis for the fetus remains above baseline in late-onset preeclampsia, the odds ratio for perinatal death and severe neonatal morbidity is only 2.0 [2,3,5]. In the presence of symptoms (headache, phosphenes, tinnitus, epigastric pain, visual changes) or alteration of laboratories (renal function tests, liver enzymes and platelet count), proteinuria can be dispensed for diagnosis [1,3].

Among its complications are: acute pulmonary edema, acute renal failure, hepatic hematoma, HELLP syndrome (hemolysis, elevated liver enzymes, and decreased platelet count), normoinsertive placental abruption, and cerebrovascular events. The latter is present in 50%–75% of maternal deaths from preeclampsia. Approximately one-third of all pregnancy-related deaths are due to complications of preeclampsia at a rate of 1.5/100,000 live births [1,5].

Some risks pertain to development of the HELLP syndrome of microangiopathic hemolysis and platelet consumption, and hepatocellular damage from periportal or focal parenchymal necrosis. Patients frequently (40%–90%) have epigastric or right upper quadrant pain. HELLP syndrome complicates 10%–20% of cases of severe preeclampsia, and develops mostly preterm (50%). In 20% of women, however, it presents in late gestation, or in 30% postpartum. HELLP without hypertension or proteinuria is reported in 10%–20% of cases. Direct complications of HELLP syndrome are abruptio placentae (9%–20%), disseminated intravascular coagulation (5%–56%), and acute renal failure (7%–36%) [5,6].

Pathophysiology

Over time, the idea that preeclampsia is only pregnancy-induced hypertension after week 20, associated with proteinuria or target organ dysfunction, has been abandoned [2,5]. We must now understand that preeclampsia is a complex syndrome, involving multiple organs, including the role of oxidative stress and endoplasmic reticulum, inflammation, secondary endothelial dysfunction, and the importance of angiogenic and antiangiogenic factors. The clinical syndrome begins with an abnormal invasion of the trophoblast long before the clinical manifestations of the disease occur [1,6,7].

During normal implantation, trophoblasts invade the decidualized endometrium, leading to remodeling of the spiral artery and obliteration of the half tunic of the spiral arteries of the myometrium, allowing increased blood flow to the placenta, all regardless of maternal vasomotor changes. In preeclampsia, trophoblasts fail to adopt an endothelial phenotype, which leads to impaired trophoblast invasion and incomplete remodeling of the spiral artery. The resulting placental ischemia leads to increased angiogenic markers, such as soluble fms-like tyrosine kinase-1 (sFlt-1) and soluble endoglin (sEng), binds to and decreases levels of vascular endothelial growth factor (VEGF) and placental growth factor (PlGF), which are important mediators of the function of endothelial cells, especially in the fenestrated endothelium (brain, liver, glomeruli) [1,8,9].

Therefore, physiopathologically, a two-stage model was developed: in the first, incomplete remodeling of the spiral artery occurs in the uterus that contributes to placental ischemia and in the second stage, the release of antiangiogenic factors from the ischemic placenta into the maternal circulation that contributes to endothelial damage consequently with hypertension and target organ damage [9].

These angiogenic markers can be used with standard clinical evaluation to help rule out proteinuric preeclampsia or preeclampsia requiring delivery within the next 7 (for sFlt-1/PlGF ratio) or 14 days (for PlGF Triage), in women with suspected preeclampsia between 20 and 34 + 6 weeks' gestation [10].

The most important risk factors for developing preeclampsia are age >40, chronic autoimmune disease, family history, type 1 diabetes mellitus, multiple gestation, nulliparity, obesity, preexisting hypertension, preexisting renal disease, previous preeclampsia, and presence of antiphospholipid antibodies [11].

Evaluation and management

Any patient diagnosed with severe preeclampsia should be admitted to an obstetric complications room or labor/delivery room. An appropriate evaluation should be performed looking for target organ damage, including a heart scan, pulmonary auscultation, a neurological examination (looking for data of focusing or altered state of consciousness), performing a fundus in case of visual changes, evaluation of urine output (Foley catheter place), and identifying data or signs of placental abruption. It is important to assess fetal well-being by cardiotocography and ultrasonography, including an estimate of fetal weight.

In addition to clinical evaluation, all patients with diagnostic criteria for preeclampsia should undergo laboratory tests including: blood count with platelet count, renal and liver function tests, 24-h urine protein measurement, and coagulation tests when platelet count is less than $100,000/mm^3$.

Upon admission, while receiving initial evaluations, the patient should receive a loading dose of magnesium sulfate ($MgSO_4$) and continue with an infusion to avoid seizures. In cases of elevated blood pressure (systolic blood pressure $\geq$160 mmHg and/or diastolic blood pressure $\geq$110 mmHg), an antihypertensive drug should be administered [12].

Once a patient with severe preeclampsia has been diagnosed, management with $MgSO_4$ should be initiated as prophylaxis for eclamptic seizures, with a loading dose of 4−6 g to be passed intravenously in 15−20 min.

Once impregnated with the charge, an infusion of $MgSO_4$ at 1−2 g/h should continue.

So far, studies have shown that $MgSO_4$ is superior to placebo and other drugs (such as diphenylhydantoin and diazepam), as a prophylactic [12−15].

During management with $MgSO_4$, the patient should be closely monitored: consciousness, osteotendinous reflexes, respiratory rate, and urine output. $MgSO_4$ infusion should be maintained for up to 24 h after termination of pregnancy. There is a risk of $MgSO_4$ poisoning during administration; signs of toxicity include: loss of osteotendinous reflexes, respiratory depression, blurred vision, and cardiotoxicity. The drug of choice to reverse the toxic effects of $MgSO_4$ is 10% calcium gluconate. The 1 g dose should be administered intravenously, and the dose repeated according to the patient's clinical response [16,17].

The goal of antihypertensive therapy is to prevent severe gestational hypertension and reduce the risk of maternal hemorrhagic stroke. The use of antihypertensives in patients who present with mild hypertension seems to reduce the rate of progression to severe hypertension in half but does not affect the incidence of developing preeclampsia [18].

The most commonly used antihypertensives for the management of hypertensive crises in preeclamptic patients are labetalol, hydralazine, and nifedipine [18].

Labetalol	Intravenous use in boluses in doses every 10−15 min progressively: 20 mg; 20 mg; 40 mg and 80 mg (maximum 300 mg/24 h).
Hydralazine	Intravenous use in boluses of 5−10 mg, every 15−20 min, up to a maximum of 30 mg.
Nifedipine	Oral use 10 mg every 20 min up to 3 doses.

Preeclampsia has its pathophysiological origin in the placenta, and is treated only with termination of pregnancy. If the fetus is <34 weeks gestation, and there is maternal stability and preserved fetal well-being, discontinuation of corticosteroids may be deferred in order to accelerate fetal lung maturation and subsequently interrupt.

Renal impact of preeclampsia

The mechanisms by which acute kidney injury (AKI) can lead to chronic kidney disease (CKD) have not been fully elucidated and the relationship between renal disease during pregnancy and hypertensive pregnancies is not well defined. AKI during pregnancy is associated with rates of maternal mortality and fetal loss that range from 30% to 60%, making it a life-threatening event. The incidence of AKI is increasing, associated with an increased risk of CKD, hypertension, and cardiovascular disease.

AKI is commonly associated with hypertensive conditions of pregnancy which themselves are associated with increased risks of cardiovascular disease later in life [19–21].

Incidences between 50% and 61% of AKI among critically ill obstetric patients with a direct correlation between disease severity and mortality [22,23]. The burden of AKI remains greater in developing countries with the World Health Organization stating that the Maternal Mortality Ratio is 239/100,000 women in developing countries compared to 12/100,000 in developed countries [24].

KDIGO defines AKI as any of the following [25,26]:

- Increase in serum creatinine by 0.3 mg/dL or more within 48 h
- Increase in serum creatinine to 1.5 times baseline or more within the last 7 days
- Urine output <0.5 mL/kg/h for 6 h.

None of these criteria are validated for use in pregnancy. There are several factors that make the diagnosis of AKI in pregnancy more challenging than in the nonpregnant state and assessment of renal function parameters used in nonpregnant individuals cannot always provide an accurate measurement in pregnancy. Due to the physiological changes and increase in glomerular filtration rate (GFR), a reduction of serum creatinine during pregnancy occurs, making the early and accurate diagnosis of AKI more difficult. This physiologic decrease in serum creatinine may mask early or mild changes in renal function. Comparison to baseline values is often not possible, as renal function parameters are often not obtained in pregnancy until injury is clinically suspected. Additionally, pregnant women may have a 30%–40% reduction in GFR without significant increases in serum creatinine [27].

The American College of Obstetricians and Gynecologists (ACOG) defines renal insufficiency in the setting of hypertensive disorders of pregnancy as a serum creatinine level >1.1 mg/dL or a doubling of the serum creatinine concentration in the absence of renal disease. However, there is no consensus on the diagnostic criteria that should be utilized for AKI in pregnancy [28].

Hypertensive disorders causing AKI by intrinsic renal can occur throughout the late second to third trimesters. Evidence suggests that hypertensive disorders of pregnancy are an important contributor to the burden of AKI in pregnancy, in particular preeclampsia and HELLP syndrome which are already associated with increased incidences of maternal and perinatal morbidity and mortality [29].

AKI has been reported to occur in 7%–60% of patients with preeclampsia when severe or when associated with HELLP syndrome. When AKI occurs, it also increases the risk of other obstetric complications. AKI in the setting of HELLP syndrome and preeclampsia has been associated with placental abruption and pulmonary edema.

AKI complicated with HELLP syndrome had placental abruption in 24% and obstetric hemorrhage in 22% [30,31]. Pregnancy complicated with preeclampsia and/or HELLP syndrome is a risk factor for chronic or future renal disease in addition to cardiovascular disease. Recent studies report that preeclampsia is associated with an increased risk of remote cardiovascular disease and end-stage renal disease (ESRD). Some studies report that renal recovery is estimated to occur in up to 74.4% of affected women [30,32].

Proteinuria has traditionally been a hallmark of preeclampsia, but it is also a nonspecific indicator of renal disease and may result from an elevated plasma protein concentration, increased glomerular permeability, decreased tubular protein reabsorption, and renal hemodynamic alterations [33].

The intact glomerular filter is almost impermeable to large proteins, and proteinuria may result from excessive permeability of the glomerular barrier for protein or impaired reabsorption of protein in the proximal tubule. With normal renal function, the amount of plasma protein entering the proximal tubule through the glomerulus depends on the glomerular plasma flow and the concentration and the filtration rate of each protein. Increased protein excretion during pregnancy is thought to be caused by the increase in GFR. The increase in GFR could result from a combination of hypervolemia and hemodilution lowering protein concentration and oncotic pressure and the increase in renal blood flow [34].

Preeclampsia was found to be associated with a distinctive glomerular appearance of endothelial vacuolization and hypertrophy of the cytoplasmic organelles. The glomeruli are enlarged and solidified (bloodless), and the swelling of the endothelial cells and to a lesser extent the mesangial cells will cause narrowing or occlusion of capillary lumens. Spargo et al. referred to these lesions by the now widely accepted term glomerular capillary endotheliosis [35,36].

Current data suggest that the number of urinary podocytes in women with preeclampsia is higher. The function of the podocyte and the slit diaphragm depends on the physiological concentrations of circulating factors, such as VEGF and its antagonist, the soluble receptor sFlt-1. In the kidney, podocytes produce VEGF, and VEGF receptors have been found in endothelial cells and podocytes themselves. Thus, paracrine and autocrine pathways could affect the integrity of the glomerular filtration barrier with tight regulation of VEGF signaling to maintain a healthy glomerulus. Women with preeclampsia have increased serum concentrations of sFlt-1 and sEng and reduced concentrations of free VEGF and free PlGF, which are proangiogenic proteins that are bound and neutralized by sFlt-1. The angiogenic imbalance in preeclampsia may play an important key role in the development of both podocyte and endothelial damage in the glomerular filtration barrier [37–40].

The presence of proteinuria was considered an essential marker for poor pregnancy outcome, and there was a tendency to consider the absence of proteinuria as reassuring when managing pregnancies complicated by hypertension. Proteinuria in combination with hypertension has long been considered to be predictive of increased maternal and neonatal adverse [41].

References

[1] Garovic VD, Dechend R, Easterling T, Karumanchi SA, Baird SMM, Magee LA, et al. Hypertension in pregnancy: diagnosis, blood pressure goals, and pharmacotherapy: a scientific statement from the American heart association. Hypertension 2022;79:E21–41. Lippincott Williams and Wilkins.
[2] Brown MA, Lindheimer MD, de Swiet M, et al. The classification and diagnosis of the hypertensive disorders of pregnancy: statement from the international society for the study of hypertension in pregnancy (ISSHP). Hypertens Pregnancy 2001;20(1):ix–xiv.
[3] American College of Obstetricians and Gynecologists, Task Force on Hypertension in Pregnancy. Hypertension in pregnancy. Report of the American College of Obstetricians and Gynecologists' task force on hypertension in pregnancy. Obstet Gynecol 2013;122(5):1122.
[4] Brown MA, Magee LA, Kenny LC, Karumanchi SA, McCarthy FP, Saito S, et al. The hypertensive disorders of pregnancy: ISSHP classification, diagnosis & management recommendations for international practice. Pregnancy Hypertens 2018;13:291–310. Elsevier B.V.
[5] Ives CW, Sinkey R, Rajapreyar I, Tita ATN, Oparil S. Preeclampsia—pathophysiology and clinical presentations: JACC state-of-the-art review. Vol. 76. J Am Coll Cardiol 2020:1690–702. Elsevier Inc.
[6] Magee LA, Nicolaides KH, von Dadelszen P, Preeclampsia, Longo DL. N Engl J Med May 12, 2022; 386(19):1817–32. Available from: http://www.nejm.org/doi/10.1056/NEJMra2109523.
[7] Roberts JM, Rich-Edwards JW, McElrath TF, Garmire L, Myatt L. Subtypes of preeclampsia: recognition and determining clinical usefulness. Hypertension 2021;77:1430–41. Lippincott Williams and Wilkins.
[8] Phipps E, Prasanna D, Brima W, Jim B. Preeclampsia: updates in pathogenesis, definitions, and guidelines. Clin J Am Soc Nephrol 2016;11:1102–13. American Society of Nephrology.
[9] Espinoza J, Vidaeff A, Pettker CM, Simhan H. Acog practice bulletin clinical management guidelines for obstetrician-gynecologists. Obstet. Gynaecol. 2019;131:e26–50. Available from: http://journals.lww.com/greenjournal.
[10] Patrocinio M, Benítez R. Guías NICE 2019 sobre tratamiento de los trastornos hipertensivos del embarazo. La visión del nefrólogo NefroPlus 2020;12(1):15–9. Vol. 12. 2020.
[11] Milne F, Redman C, Walker J, et al. The pre-eclampsia community guideline (PRECOG): how to screen for and detect onset of preeclampsia in the community. BMJ 2005;330:576–80.
[12] Magee L, von Dadelszen P. Prevention and treatment of postpartum hypertension. Cochrane Database Syst Rev 2013;(4):CD004351.
[13] Duley L, Henderson-Smart DJ, Chou D. Magnesium sulphate versus phenytoin for eclampsia. Cochrane Database Syst Rev 2010;10:CD000128.
[14] Duley L, Henderson-Smart DJ, Walker G, et al. Magnesium sulphate versus diazepam for eclampsia. Cochrane Database Syst Rev 2010;12:CD000127.
[15] Belfort MA, Anthony J, Saade GR, et al. A comparison of magnesium sulfate and nimodipine for the prevention of eclampsia. N Engl J Med 2003;348(4):304–11.
[16] Euser AG, Cipolla MJ. Resistance artery vasodilation to magnesium sulfate during pregnancy and the postpartum state. Am J Physiol Heart Circ Physiol 2005;288(4):H1521–5.

[17] Hatab MR, Zeeman GG, Twickler DM. The effect of magnesium sulfate on large cerebral artery blood flow in preeclampsia. J Matern Fetal Neonatal Med 2005;17(3):187–92.
[18] Abalos E, Duley L, Steyn D. Antihypertensive drug therapy for mild to moderate hypertension during pregnancy [review]. Cochrane Database Syst Rev 2014;(2):CD002252.
[19] Bentata Y, Housni B, Mimouni A, Azzourzi A, Abouqal R. Acute kidney injury related to pregnancy in developing countries: etiology and risk factors in an intensive care unit. J Nephrol. 2012;25: 764–75.
[20] Silver S, Siew E. Follow-up care in acute kidney injury: lost in transition. Adv Chronic Kidney Dis 2017;24:246–52.
[21] Parr S, Matheny M, Abdel-Kader K, Greevy Jr R, Bian A, Fly J, et al. Acute kidney injury is a risk factor for subsequent proteinuria. Kidney Int 2018;93:460–9.
[22] Tyagi A, Luthra A, Kumar M, Das S. Epidemiology of acute kidney injury and the role of urinary TIMP-2 and IGFBP7: a prospective cohort study in critically ill obstetric patients. Int J Obst Anesthesia 2018;36:77–84.
[23] Bedford M, Stevens P, Wheeler T, Farmer C. What is the real impact of acute kidney injury? BMC Nephrol 2014;15:1–9.
[24] WHO, UNICEF, UNFPA, Group WB, Division UNP. Trends in maternal mortality: 1990 to 2015 estimates by WHO, UNICEF, UNFPA, World Bank Group and the United Nations Population Division. Geneva. 2015.
[25] Lopes J, Jorge S. The RIFLE and AKIN classifications for acute kidney injury: a critical and comprehensive review. Clin Kidney J 2013;6:8–14.
[26] Khwaja A. KDIGO clinical practice guidelines for acute kidney injury. Nephron Clin Pract 2012;120: c179–84.
[27] Fakhouri F, Vercel C, Fremeaux-Bacchi V. Obstetric nephrology: AKI and thrombotic microangiopathies in pregnancy. Clin J Am Soc Nephrol 2012;7:2100–6. https://doi.org/10.2215/CJN.13121211.
[28] ACOG Practice Bulletin No. 202: gestational hypertension and preeclampsia. Obstet Gynecol 2019; 133:e1–25.
[29] Liu Y, Ma X, Zheng J, Liu X, Yan T. Pregnancy outcomes in patients with acute kidney injury during pregnancy: a systematic review and meta-analysis. BMC Pregnancy Childbirth 2017;17:235.
[30] Ye W, Shu H, Yu Y, Li H, Chen L, Liu J, et al. Acute kidney injury in patients with HELLP syndrome. Int Urol Nephrol 2019;51:1199–206.
[31] Gul A, Aslan H, Cebeci A, Polat I, Ulusoy S, Ceylan Y. Maternal and fetal outcomes in HELLP syndrome complicated with acute renal failure. Ren Fail 2004;26:557–62.
[32] Vikse BE, Irgens LM, Leivestad T, Skjaerven R, Iversen BM. Preeclampsia and the risk of end-stage renal disease. N Engl J Med 2008;359:800–9.
[33] Fishel Bartal M, Lindheimer MD, Sibai BM. Proteinuria during pregnancy: definition, pathophysiology, methodology, and clinical significance. Am J Obstet Gynecol 2022;226(2S):s819–34. https://doi.org/10.1016/j.ajog.2020.08.108. Epub 2020 Sep 1. PMID: 32882208.
[34] Bökenkamp A. Proteinuria-take a closer look. Pediatr Nephrol 2020;35:533–41.
[35] Gaber LW, Spargo BH, Lindheimer MD. Renal pathology in pre-eclampsia. Baillieres Clin Obstet Gynaecol 1987;1:971–95. 33. Spargo BH0, Lichtig C, Luger AM, Katz AI.
[36] Lindheimer MD. The renal lesion in preeclampsia. Perspect Nephrol Hypertens 1976;5:129–37.
[37] Henao DE, Mathieson PW, Saleem MA, Bueno JC, Cadavid A. A novel renal perspective of preeclampsia: a look from the podocyte. Nephrol Dial Transplant 2007;22:1477.
[38] Henao DE, Saleem MA, Cadavid AP. Glomerular disturbances in preeclampsia:disruption between glomerular endothelium and podocyte symbiosis. Hypertens Pregnancy 2010;29:10–20.
[39] Foster RR, Hole R, Anderson K, et al. Functional evidence that vascular endothelial growth factor may act as an autocrine factor on human podocytes. Am J Physiol Renal Physiol 2003;284:F1263–73.
[40] Maynard SE, Min JY, Merchan J, et al. Excess placental soluble fms-like tyrosine kinase 1 (sFlt1) may contribute to endothelial dysfunction, hypertension, and proteinuria in preeclampsia. J Clin Invest 2003;111:649–58.
[41] Thangaratinam S, Coomarasamy A, O'Mahony F, et al. Estimation of proteinuria as a predictor of complications of pre-eclampsia: a systematic review. BMC Med 2009;7:10.

CHAPTER 13

Glomerular diseases in pregnancy

Ahsan Ahmed[1], Anirban Bose[2] and Md Mostafizur Rahman Akash[3]
[1]Critical Care Medicine, KPC Medical College and Hospital, Kolkata, West Bengal, India; [2]Medica Institute of Critical Care, Medica Superspecialty Hospital, Kolkata, West Bengal, India; [3]Shaheed Monsur Ali Medical College Hospital, Uttara, Dhaka, Bangladesh

Introduction

In the past, women with glomerulonephritis rarely get conceived due to fear of progression of the disease and adverse maternofetal outcome. Today 0.7% women of childbearing age is suffering from chronic kidney disease (CKD) stage 3%–5% and 3% women of same age group suffering from CKD stage 1–2 [1]. With advancement of medical science, pregnancy became a feasible option for more women with glomerulonephritis. Exact prevalence of glomerular disease (GD) in pregnant patient is unknown due to lack of routine renal function testing in pregnancy. Early GD can be missed in entire period of pregnancy though a greater number of patients with mild GD reported more recently due to advanced maternal age with higher incidence of another comorbidity. Although all types of GDs are found in pregnancy, but focal segmental glomerulosclerosis (FSGS), immunoglobin A (IgA) nephropathy, and lupus nephritis (LN) are among the commonest of all.

Effects of pregnancy on glomerular disease

The stage of renal disease at the time of conception is a major determinant of disease progression. Data are spares, more precisely disease specific and understanding is severely limited due to sample size, heterogenicity, reliance on serum creatinine, and methods of glomerular filtration rate (GFR) estimation.

Women with GD with preserved renal function usually do not have accelerated disease progression during pregnancy period irrespective of etiology [2]. On the other hand, women with more advanced CKD likely to suffer from more accelerated disease progression during pregnancy [3] although data are spares and limited.

Effects of glomerular disease on pregnancy

Pregnancy outcomes including but not limited to abortion, still birth, perinatal death, small for gestational age, preterm delivery, malformation, hereditary kidney disease, CKD-hypertension, and long-term complex neurological disease depend upon disease

The Kidney of the Critically Ill Pregnant Woman
ISBN 978-0-443-21473-8,
https://doi.org/10.1016/B978-0-443-21473-8.00013-6

activity, severity of renal impairment, proteinuria, and hypertension. Risk of maternal death is also increased in women with advanced CKD [4].

Diagnosis

No clear definition available for nephrotic syndrome in pregnancy. Pregnant patients with proteinuria and/or subacute/acute kidney injury usually suspected for GD. History, clinical examinations (searching for systemic manifestation of primary disorder), blood investigations for autoantibodies and compliments, and kidney biopsy have major role to reach the diagnosis.

Kidney biopsy rarely required as incidence of de novo GD is very low and availability of alternate diagnostic tools. Kidney biopsy only indicated when diagnosis can not be made by alternative tests and that diagnosis will change the course of the treatment. Historically kidney biopsy was associated with significant risks, but now a days, it is considered safe in experienced hand [5], more if done in early pregnancy.

Distinguishing preeclampsia from worsening of GD is difficult due to its frequent association with renal diseases and heterogeneity. Soluble fms-like tyrosine kinase 1/placental growth factor ratio showing promise to differentiate between two conditions (elevated in preeclampsia) but need more data to establish its role [6].

Management of individual disease

Lupus nephritis

Women with LN need prepregnancy counseling. Methotrexate and mycophenolate should be stopped or switched to azathioprine 6 weeks before conception [7]. Hydroxychloroquine has favorable benefit over risk ratio and should be considered in all pregnant patient with LN [8]. Steroids and calcineurin inhibitors are safe in pregnancy [9]. Rituximab can be considered in severe and resistant cases. Low dose aspirin should be prescribed to all patient with LN and in addition to heparin in patients with antiphospholipid antibodies [9]. Female with anti-Ro/anti-La antibodies have higher chance of fetal atrioventricular block, need fetal heart rhythm monitoring periodically but recently questioned [10].

Flare is a major concern during pregnancy. Reduced complement level and increased proteinuria and/or serum creatinine are suggestive of flare and renal biopsy is rarely required. Steroids along with azathioprine or tacrolimus is the mainstay of treatment [7]. Rituximab can be used in severe flair [7]. In organ or life-threatening flare, cyclophosphamide can be used although it is contraindicated in pregnancy [11]. Mycophenolate can be used in late pregnancy when chance of teratogenicity is no longer present [7].

IgA nephropathy

IgA nephropathy is the most common primary GD. The course of pregnancy usually remains uneventful. ACE inhibitor, which is the main stay of the treatment needs to stop before pregnancy and should restart in postpartum period. Treatment is usually supportive in pregnancy and rarely required steroids [12]. Severe disease progression and flare requiring immunosuppression is rare in IgA nephropathy with pregnancies.

Membranous nephropathy

Pregnancy in membranous nephropathy patients shows adverse maternal—fetal outcome [13]. Steroid or tacrolimus can be used to treat membranous nephropathy [12]. Rituximab can be used before or early in the pregnancy. As high rate of venous thromboembolism is associated with membranous nephropathy, low molecular weight heparin should be used during pregnancy and breastfeeding [14]. Transplacental transfer of PLA2R antibodies was evident without any major clinical significance. PLA2R antibodies also found in breast milk [4]. Neonatal membranous nephropathy caused by anti-neutral endopeptidase antibodies, but mother remains unaffected due to lack of target antigen. Majority of neonatal membranous nephropathy are transient though some neonate may develop severe features requiring plasmapheresis [15].

Focal segmental glomerulosclerosis

FSGS is commonly found in biopsies done antenatally, but overall data are spares. FSGS is commonly associated with adverse pregnancy outcomes [16]. Steroids or tacrolimus remains the main stay of treatment [12].

Minimal change disease

Relapse of minimal change disease after remission in pregnancy is unusual. Relapse is commonly associated with corticosteroids-dependent minimal change disease patients. Relapse usually treated with bolus steroids followed by oral maintenance steroid therapy [4]. Calcineurin inhibitors are feasible alternative. Rituximab can be used in minimal change disease with high recurrence rate [17].

Vasculitis in pregnancy

Although vasculitis usually affects old age people, women in child-bearing age can also suffer from vasculitis. Mainly two types of kidney vasculitis involved namely antineutrophil cytoplasmic antibody vasculitis and IgA vasculitis. Disease flair is much common, predominantly mild to moderate, but severe flair is also reported. Maternal—fetal adverse outcome is also common in vasculitis. Severe flair during pregnancy is associated with poor maternal fetal outcome, while remission well before pregnancy is usually associated with good outcome. Treatment consists of either steroids or azathioprine. Rituximab is also used as a maintenance therapy. Rapidly progressing disease can be treated with steroids, rituximab, and/or plasma exchange [18]. Cyclophosphamide is usually contraindicated but can be used in organ or life-threatening condition [18]. Intravenous immunoglobulin is also used in this situation [18]. Transplacental transfer of antimyeloperoxidase-antineutrophil cytoplasmic antibody causing neonatal alveolar hemorrhage and acute kidney injury requires immunosuppressive therapies and exchange transfusion [19].

General care

Blood pressure management is the most important aspect of care. There is no specific blood pressure goal in pregnant women, but general consensus is maintained below 140/90 mm of Hg. Methyldopa, labetalol, hydralazine, and long acting nifedipine are safe to use. RAS inhibitors are strictly contraindicated for its teratogenic effects. Diuretics are usually avoided in view of risk of volume contraction and acute kidney injury.

Peripheral edema is very common in pregnancy and usually managed conservatively with compression stockings and avoiding prolonged standing. Diuretics are rarely used when conservative management failed. Role of albumin infusion is controversial.

Pregnancy with GD is a hypercoaguable state. Thromboprophylaxis is indicated when serum albumin level is less than 2 g/dL or there is presence of other risk factors like obesity, immobility, and membranous glomerulonephritis. Subcutaneous low molecular weight heparin is the drug of choice as it does not cross placenta. Low molecular weight heparin should be stopped prior to anticipated delivery and restart as soon as possible postpartum.

Dyslipidemia is common in pregnancy with GD. Statins, fibrates, ezetimibe, and niacins are considered unsafe in pregnancy and bile acid sequestrants considered safe in pregnancy.

Low protein diet along with supplementation of amino acids and ketoacids can be considered in absence of other antiproteinuric treatment.

Delivery

Timing and mode of delivery should be guided by obstetric factors and irrespective of types and severity of glomerulonephritis itself. However, opinion of nephrologist should be taken wherever available for best maternofetal outcome without worsening GD.

Postpartum

Immunosuppressants and low molecular weight heparin should be continued in the postpartum period. Enalapril can be restarted. There are no contraindications in using enalapril for breast feedings. These drugs are safe during breast feeding. Follow up with nephrologist needed and if renal biopsy is required, it should be done at 4–6 weeks post-partum period if there is no clinical urgency. Regular blood pressure monitoring is needed and adequate control is utmost necessary [20].

References

[1] Williams D, Davidson J. Chronic kidney disease in pregnancy. BMJ 2008;336:211–5.

[2] Jungers P, Houillier P, Forget D, et al. Influence of pregnancy on the course of primary chronic glomerulonephritis. Lancet 1995;346:1122–4.

[3] Jones DC, Hayslett JP. Outcome of pregnancy in women with moderate to severe renal insufficiency. N Engl J Med 1996;335:226–32.

[4] Fakhouri F, Schwotzer N, Cabiddu G, et al. Glomerular disease in pregnancy: pragmatic recommendations for clinical management. Kidney Int 2023;103(2):264–81.

[5] Chen TK, Gelber AC, Witter FR, et al. Renal biopsy in the management of lupus nephritis during pregnancy. Lupus 2015;24:147–54.

[6] Rolfo A, Attini R, Tavassoli E, et al. Is it possible to differentiate chronic kidney disease and pre-eclampsia by means of new and old biomarkers? A prospective study. Dis Markers 2015;2015:127083.

[7] Lightstone L, Hladunewich MA. Lupus nephritis and pregnancy: concerns and management. Semin Nephrol 2017;37(4):347–53.

[8] Kaplan YC, Ozsarfati J, Nickel C, et al. Reproductive outcomes following hydroxychloroquine use autoimmune disease: a systematic review and meta-analysis. Br J Clin Pharmacol 2016;81:835–48.

[9] Andreoli L, Bertsias GK, Agmon-Levin N, et al. EULAR recommendations for women's health and the management of family planning, assisted reproduction, pregnancy and menopause in patients with systemic lupus erythematous and/or antiphospholipid syndrome. Ann Rheum Dis 2017;76:476–85.

[10] Costedoat-Chalumeau NM, Fischer-Betz R, Levesque K, et al. Routine repeated echocardiography monitoring of fetuses exposed to maternal anti-SSA antibodies: time to question the dogma. Lancet Rheumatol 2019;1:e187–93.

[11] Lannes G, Elias FR, Cunha B, et al. Successful pregnancy after cyclophosphamide therapy for lupus nephritis. Arch Gynecol Obstet 2011;283(1):61–5.

[12] Gleeson S, Lightstone L. Glomerular disease and pregnancy. Adv Chronic Kidney Dis 2020;27(6): 469–76.

[13] Liu ZN, Cui Z, He YD, et al. Membranous nephropathy in pregnancy. Am J Nephrol 2020;51: 304–17.

[14] Bates SM, Rajsekhar A, Middeldrorp S, et al. American Society of Hematology 2018 guidelines or management of venous thromboembolism: venous thromboembolism in the context of pregnancy. Blood Adv 2018;2(22):3317–59.

[15] Debiec H, Guigonis V, Mougenot B, et al. Antenatal membranous glomerulonephritis due to anti-neutral endopeptidase. N Engl J Med 2002;346:2053–60.

[16] O'Shaughnessy MM, Jobson MA, Sims K, et al. Pregnancy outcomes in patients with glomerular disease attending in a single academic centre in North Carolina. Am J Nephro 2017;45(5):442–51.

[17] Holden F, Bramham K, Clark K, et al. Rituximab for the maintenance of minimal change nephropathy – a report of two pregnancies. Obstet Med 2020;13:145–7.

[18] Veltri NL, Hladunewich M, Bhasin A, et al. De novo antineutrophil cytoplasmic antibody-associated vasculitis in pregnancy: a systemic review on maternal, pregnancy and fetal outcomes. Clin Kidney J 2018;11(5):659–66.

[19] Bansal PJ, Tobin MC. Neonatal microscopic polyangitis secondary to transfer of maternal myeloperoxidase-antineutrophilic cytoplasmic antibody resulting in neonatal pulmonary haemorrhage and renal involvement. Ann Allergy Asthma Immunol 2004;93:398–401.

[20] Day C, Hewins P, Hildebrand S, et al. The role of renal biopsy in women with kidney disease identified in pregnancy. Nephrol Dial Transplant 2008;23:201–6.

CHAPTER 14

Renal biopsies during pregnancy

Jamshid Roozbeh and Fatemeh Masjedi
Shiraz Nephro-Urology Research Center, Shiraz University of Medical Sciences, Shiraz, Iran

Introduction

There are several unresolved matters regarding the performance of kidney biopsy in pregnant women [1], including determining appropriate indications for biopsy, such as a sudden decline in kidney function before 32 weeks of gestation and differentiation from preeclampsia (PEC) and the efficacy of empirical therapy, particularly with steroids [2—7]. Some researchers have proposed delaying kidney biopsy until at least 2 months after birth due to the fast healing of preeclamptic lesions. On the other hand, others have emphasized the need to promptly identify and intervene in cases of various glomerular disorders [8,9].

There are numerous compelling reasons to carefully examine the benefits and risks of renal biopsy during pregnancy. Improvements in maternal and fetal care allow infants delivered as early as 24 or 25 weeks of gestation to survive without serious complications, shifting the balance between the dangers of early delivery and invasive interventions [10,11]. Conversely, there is a growing emphasis on the benefits of early detection of glomerular diseases. However, there is uncertainty about the applicability of this phenomenon to pregnant women since previous studies have proposed the possibility of hyperfiltration-induced damage playing an additional role in this population [12,13].

The ethical considerations are also undergoing transformation. In a context where decisions are made collectively, kidney biopsy may be contemplated under certain circumstances, such as during a flare-up of systemic lupus erythematosus (SLE) or when there is a rapid decline in kidney function accompanied by favorable fetal indicators. The objective would be to ascertain potential treatments that could enable the continuation of the pregnancy until a relatively safe stage is reached, ensuring positive fetal outcomes [14—16].

Furthermore, doing a kidney biopsy during pregnancy presents several unique challenges. The feasibility of using the conventional prone stance for biopsy is sometimes limited in advanced stages of pregnancy, and adopting other positions, such as lateral decubitus or sitting, may provide additional technical challenges to the procedure [17]. Pregnant women being evaluated for kidney biopsy may have an increased likelihood of experiencing maternal or fetal difficulties associated with the biopsy procedure due to their preexisting comorbidities [18]. Serious issues, such as bleeding, may impact the mother and fetus. However, it is worth noting that a biopsy has the potential to provide a conclusive diagnosis and subsequently result in a modification of the treatment approach [19]. Hence,

The Kidney of the Critically Ill Pregnant Woman
ISBN 978-0-443-21473-8
https://doi.org/10.1016/B978-0-443-21473-8.00014-8

it is essential to thoroughly evaluate the advantages of kidney biopsy in terms of clinical diagnosis and therapy in pregnancy while also considering the possible risks it may provide. The existing body of research on the safety of renal biopsy during pregnancy mostly consists of studies conducted at single centers. The limitations of these studies include small sample size, the use of an observational design, variations in the rationale for kidney biopsy, and discrepancies in the definitions of severe postbiopsy sequelae [20].

Therefore, this issue remains controversial despite extensive investigation, and there is no consensus regarding indications, contraindications, timing, value, or safety. In this chapter, we review the indications and risk/benefit ratio for renal biopsy during pregnancy over the past 20 years.

Renal biopsy indications during pregnancy

A biopsy may be recommended in pregnant individuals when there is a rapid and significant decline in glomerular filtration rate (GFR) or the onset of nephrotic-range proteinuria. This diagnostic procedure is considered when other less invasive methods have been unsuccessful in determining the underlying cause of the condition [17]. Kidney biopsy has significant importance in diagnosing de novo lupus nephritis or acute lupus flare in individuals who have previously been diagnosed with lupus nephritis. The diagnosis of lupus nephritis cannot be solely based on clinical suspicion, even when laboratory assessment provides supporting evidence. This is due to the critical importance of underlying histopathology in directing appropriate therapy [21,22].

Chen et al. [23] investigated kidney biopsy outcomes in a cohort of 11 female individuals diagnosed with SLE at Johns Hopkins University over 11 years from 2001 to 2012. The decision to do a kidney biopsy was made based on the clinical assessment of the attending rheumatologist and nephrologist. The patients had biopsies while in the prone position, with a median gestational age of 16 weeks (range 9–27 weeks), under the direction of ultrasonography. To meet the criteria for kidney biopsy, it was required that all patients exhibited a blood pressure equal to or less than 160/100 mm Hg, a platelet count of at least 50,000, a partial thromboplastin time below 42.1 s, and an international normalized ratio lower than 1.3. All women in this work had a minimum of 500 mg/24 h proteinuria; 73% displayed positive antidouble-stranded DNA titer, whereas 82% had low complement levels. The biopsy findings led to a modification in the clinical care of the condition in 91% (10 out of 11) of the patients. The research findings indicate that the advantages of doing a kidney biopsy in pregnant individuals diagnosed with lupus nephritis outweigh the potential drawbacks. This is particularly crucial considering that leaving the condition untreated may lead to substantial and long-lasting organ damage in the mother, as well as unfavorable pregnancy outcomes [24].

Nevertheless, the assessment of the suitability of kidney biopsy in pregnant individuals is contingent upon clinical discretion, the specific clinical condition of the patient, the stage of pregnancy, the likelihood of adverse effects from the procedure compared to the potential advantages of obtaining a conclusive diagnosis, and the potential impact of the biopsy on the overall clinical treatment plan.

Renal biopsies risk/benefit ratio during pregnancy

The fundamental rationale for doing a kidney biopsy during pregnancy is based on the expectation that the obtained data would significantly impact the subsequent treatment plan and/or provide valuable prognostic insights. The use of a biopsy may aid in the determination of whether it is necessary to terminate a pregnancy or initiate an early birth, ensuring that neither the viability of the fetus nor the well-being of the mother is compromised. Alternatively, individuals may choose a cautious management approach throughout pregnancy and do a postpartum biopsy afterward. Significant variability has been seen across many studies regarding the criteria for biopsy. Due to the varying objectives of studies, the evolution of biopsy methods, and the absence of a universally accepted method for evaluating the risk/benefit ratio, divergent viewpoints have emerged about the appropriateness of performing biopsies during pregnancy. Hence, using a kidney biopsy during pregnancy is often warranted in cases where there is a gradual decline in kidney function and/or the presence of significant proteinuria, as these factors may potentially impact the pregnancy outcomes. Additionally, the performance of a kidney biopsy is necessary to establish a definitive diagnosis, which in turn is crucial for determining the appropriate course of therapy [20,23,25–28] (Table 14.1).

Table 14.1 Considerations for using kidney biopsy in pregnant women.

(1)	Most clinicians hesitate to do a kidney biopsy during pregnancy. This is due to the increased risk of problems, estimated to be as high as 7% in a systematic review, compared to 1% after birth (2% risk of significant bleeding in the late second trimester) [20]. The risk of bleeding has been linked to increased renal blood flow, which is considered reversible within 3 months after birth [29].
(2)	More recent studies showed that the risks of a kidney biopsy during pregnancy are minimal when carried out by an experienced specialist and that this treatment may be more routinely considered [23,26].
(3)	The use of laboratory tests, such as the assessment of proteinuria, the examination of antibodies for LN, ANCA, and anti-PLA2R antibodies, might potentially result in the first administration of empirical medication, with the biopsy being deferred until after childbirth [29].
(4)	When contemplating a kidney biopsy, timing is critical. The kidney biopsy risks are generally modest in early pregnancy (<12 weeks), and the benefits of accurately identifying the kidney illness are significant. This risk profile varies during pregnancy and may decrease between 30 and 34 weeks [29].

ANCA, antineutrophil cytoplasm antibody; *LN*, lupus nephritis; *PLA2R*, phospholipase A2 receptor.

A retrospective research conducted in 2008 examined a cohort of 95 women who had renal illness during pregnancy for 20 years [19]. Given the duration of the trial, there were variations in the approach to kidney biopsy. The first biopsies were conducted using a 14-gauge Tru-Cut needle after the localization of the kidney by ultrasound imaging. In contrast, real-time ultrasound localization and automated 16G or 18G biopsy needles were used for later biopsies. Out of the whole cohort of 95 women, a subset of 20 individuals had a serious medical condition that justified the need for an antepartum biopsy. The median gestational age at the moment of biopsy was determined to be 20.5 weeks. A total of two biopsies were conducted during the third trimester, while an additional 75 biopsies were performed after childbirth. Among the women who had a biopsy during pregnancy, it was observed that a single patient exhibited mild hematuria. No problems were reported in the biopsied postpartum cases. The study primarily examined the criteria for performing biopsies on 20 antepartum women. Out of these women, four had a prior history of SLE, with renal disease manifesting solely during pregnancy. The remaining 16 women exhibited either deteriorating levels of proteinuria and hypertension, nephrotic syndrome accompanied by proteinuria and impaired renal function, or positive autoimmune serologies combined with proteinuria. Out of the total sample size of 20 individuals, 7 were found to have clinical PEC. The biopsies revealed various abnormalities, including focal and segmental glomerulosclerosis (FSGS), lupus nephritis, IgA nephropathy, interstitial nephritis, and several other conditions. The clinical consequences were noteworthy: the identification of conditions in 9 out of the 20 biopsies conducted before childbirth resulted in a modification of the treatment approach. Further monitoring revealed that 6 out of the 20 patients developed end-stage renal disease, compared to 6 out of the 75 patients who had biopsies after childbirth. The observed variation might be attributed to the severity of the renal disease that initially necessitated the antepartum biopsy. The authors propose that doing a biopsy before 28 weeks of gestation is advisable in pregnant women who meet certain criteria since this procedure might potentially result in therapeutic intervention, enhancing the likelihood of infant survival. The authors provided comprehensive instructions for conducting biopsies throughout the first and second trimesters. Additionally, they discouraged the regular use of biopsy to differentiate between PEC and other renal disorders. However, it is worth noting that this recommendation seems to rely on clinical variables rather than histopathologic characteristics or prior literature studies.

In 2013, a group of Italian researchers published a systematic literature analysis on renal biopsy in pregnancy [20]. A total of 39 papers were evaluated, with the majority of them focusing on postpartum biopsy descriptions. As previously mentioned, the results and interpretations were heterogeneous, indicating variations in the study design, the biopsy procedure utilized, and the characteristics of the individuals included in the research. Based on the obtained data, the authors arrived at the following conclusions: (1) The limited number of cases in most studies reflects the disinclination to doing kidney biopsies during pregnancy, (2) the previous conclusion that performing a biopsy during pregnancy is a "morbid" procedure with a high complication rate is reaffirmed, along with the recommendation that it should only be performed when emergent therapy is required, (3) the risk is greatest in the "gray area" between nonviability and viability of the fetus, (4) in addition, there is a higher incidence of complications associated with biopsies performed during pregnancy compared to the postpartum period (7% vs. 1%), and (5) the authors concluded that despite being a potentially dangerous procedure, biopsies have numerous advantages. In 66% of cases, there was a change in treatment based on the results of a biopsy, which may increase its availability as novel diagnostic modalities.

Another study conducted in 2015 examined the utility of renal biopsy in pregnant individuals diagnosed with lupus nephritis [23]. The research included a cohort of 11 individuals who had biopsy procedures over 11 years. All participants met the diagnostic criteria for SLE as outlined by the American College of Rheumatology (ACR) [30]. Eight women were diagnosed with SLE before pregnancy, while the other three had signs and symptoms throughout their pregnancies. Among the 11 individuals, it was found that two of them had a preexisting medical condition known as lupus nephritis. All individuals included in the study were in the first or second trimester of pregnancy. The biopsies were conducted with 18-gauge needles with real-time ultrasound guidance. No maternal problems resulting from the biopsy technique were recorded. The histopathological analysis revealed evidence of proliferative lupus nephritis (ISN/RPS class III or IV) in 10 of 11 patients. Furthermore, an extra or isolated occurrence of membrane lupus nephritis (class V) was seen in 6 individuals. The treatment plan was modified in 10 out of the 11 patients based on the results obtained from the biopsy. This modification included the early terminations in 2 cases. Three other individuals had premature termination of pregnancy, either with or without fetal death, within 3 weeks after the biopsy procedure. In all cases, these conditions occurred for clinical reasons and were not related to the biopsy procedure. The remaining six pregnancies culminated in the successful birth of live infants between 27 and 38 weeks of gestation. The authors concluded that doing renal biopsy during pregnancy in individuals who meet the criteria for lupus is crucial to rule out other kidney diseases and accurately identify lupus nephritis in cases where it is present. This enables appropriate adjustments to therapy, if necessary. The authors asserted that the selection criteria for biopsy in pregnant patients should align with those used for nonpregnant individuals. They emphasized that biopsy procedures should not be conducted beyond the 28th week of gestation.

A review (2016) on optimal approaches to performing native kidney biopsies [31] incorporates a segment on pregnancy. This review suggests that circumstances warranting renal biopsy during pregnancy should encompass unexplained renal failure, symptomatic nephrotic syndrome, guidance in managing lupus nephritis, and the diagnosis or exclusion of PEC.

While previous studies have shown a higher likelihood of bleeding, more current evidence indicates that the procedure may be considered safe when performed by skilled professionals [23,26]. Nevertheless, the utilization of kidney biopsy in the context of pregnancy is infrequent in clinical settings. This can be attributed to the low occurrence of de novo glomerular diseases, the presence of alternative diagnostic methods, and delayed referral. In late gestation, inducing delivery may be considered a viable alternative, particularly in well-equipped settings with readily accessible specialized perinatal care and no associated costs [32]. For example, the only series published in the last 5 years, from the largest referral center for complicated pregnancy in Mexico, includes only 20 kidney biopsies conducted over 5 years [33].

Conclusion

Patients with glomerular disorders are more likely to have difficulties during pregnancy, and it may be challenging to distinguish between the reactivation or onset of the disease and pregnancy-related symptoms such as PEC. A suitable diagnosis is critical for the mother- and fetus-related outcomes. The use of renal biopsy during pregnancy continues to be controversial. Although several authors argue for the safety of the technique, assessing the risk/benefit ratio is crucial when considering its implementation. This assessment involves carefully evaluating the potential risks to maternal health and the overall result of the pregnancy. It is important to consider the potential for increased risk of bleeding in women who present with PEC or uncontrolled hypertension. If there is a belief that a kidney biopsy might lead to a prospective alteration in therapeutic intervention, it may be defensible to perform this procedure. In this context, a kidney biopsy is very useful for diagnosing patients and determining the appropriate course of treatment. Based on the available literature and our personal experience, it is advisable to consider doing a kidney biopsy in individuals exhibiting urine irregularities or impaired renal function prior to pregnancy planning. It may be beneficial in determining the optimal period for conception. The use of preconception counseling and pregnancy planning in individuals with inactive disease, stable renal function, and normal blood pressure can potentially mitigate problems during pregnancy. It is recommended that healthcare professionals contemplate performing a kidney biopsy during pregnancy, preferably before the 25th week of gestation. This course of action aims to mitigate the likelihood of complications in individuals exhibiting active urinary sediment, substantial proteinuria (exceeding 500 mg per day), or a decline in GFR when less invasive diagnostic methods have proven insufficient in determining the underlying cause of the disease. Furthermore, it is advisable to have postpartum renal evaluations, which may include a kidney biopsy, for individuals who experience changes in renal function or ongoing proteinuria after childbirth (Table 14.2).

Table 14.2 Recommendations for renal biopsy in pregnant women diagnosed with glomerular disease.

Kidney biopsy before pregnancy
Perform a kidney biopsy in individuals with unexplained proteinuria, hematuria, or deterioration in kidney function before planning a pregnancy: - To determine the best therapeutic strategy before pregnancy; - To determine the optimal time of conception during stable remission of the underlying disease to reduce the risk of pregnancy complications; - To predict the pregnancy outcomes and diminish maternal and fetal risks.
Kidney biopsy during pregnancy
Consider a kidney biopsy in pregnant individuals with a rapid decline in glomerular filtration rate (GFR), active urinary sediment, or proteinuria >500 mg/day and have failed to clarify disease etiology using less invasive diagnostic testing. - A kidney biopsy during pregnancy is essential to distinguish glomerular disease that first manifests during pregnancy, reactivation of the underlying disease (more common in patients with lupus nephritis), or preeclampsia; - Reduce bleeding risks by monitoring coagulation factors and platelet count (especially in cases of severe thrombocytopenia), lowering blood pressure with antihypertensives, and discontinuing aspirin for at least 1 week before the procedure; - Avoid performing a biopsy after 28 weeks of gestation to balance the procedure's risks with the possibility of premature birth.
Postpartum kidney biopsy
Consider a kidney biopsy in individuals with kidney damage during pregnancy, especially if they have persistent or rising proteinuria after birth, to enhance long-term health outcomes. - Patients with lupus nephritis may have a higher risk of renal flares within 3 months after birth, necessitating a kidney biopsy for proper diagnosis and care.

References

[1] Petrucco OM, Thomson NM, Lawrence JR, Weldon MW. Immunofluorescent studies in renal biopsies in pre-eclampsia. Br Med J 1974;1(5906):473–6. https://doi.org/10.1136/bmj.1.5906.473.

[2] Lindheimer MD, Davison JM. Renal biopsy during pregnancy: 'to b ... or not to b ...?' Br J Obstet Gynaecol 1987;94(10):932–4. https://doi.org/10.1111/j.1471-0528.1987.tb02265.x.

[3] Stillman IE, Karumanchi SA. The glomerular injury of preeclampsia. J Am Soc Nephrol 2007;18(8): 2281–4. https://doi.org/10.1681/asn.2007020255.

[4] Lindheimer MD, Mahowald MB. Glomerular endotheliosis in normal pregnancy and pre-eclampsia. Bjog 2004;111(2):191. https://doi.org/10.1046/j.1471-0528.2003.00073.x. author reply 3-5; discussion 5.

[5] Pandya BK, Gibson SP, Robertson IG. Nephrotic syndrome in early pregnancy–is renal biopsy always necessary? Nephrol Dial Transplant 2002;17(4):672–4. https://doi.org/10.1093/ndt/17.4.672.

[6] Piccoli GB, Conijn A, Attini R, Biolcati M, Bossotti C, Consiglio V, et al. Pregnancy in chronic kidney disease: need for a common language. J Nephrol 2011;24(3):282–99. https://doi.org/10.5301/jn.2011.7978.

[7] Williams D, Davison J. Chronic kidney disease in pregnancy. Bmj 2008;336(7637):211–5. https://doi.org/10.1136/bmj.39406.652986.BE.

[8] Steegers EA, von Dadelszen P, Duvekot JJ, Pijnenborg R. Pre-eclampsia. Lancet 2010;376(9741): 631–44. https://doi.org/10.1016/s0140-6736(10)60279-6.
[9] Hladunewich M, Karumanchi SA, Lafayette R. Pathophysiology of the clinical manifestations of preeclampsia. Clin J Am Soc Nephrol 2007;2(3):543–9. https://doi.org/10.2215/cjn.03761106.
[10] Guillen U, DeMauro S, Ma L, Zupancic J, Wang E, Gafni A, et al. Survival rates in extremely low birthweight infants depend on the denominator: avoiding potential for bias by specifying denominators. Am J Obstet Gynecol 2011;205(4):329.e1–7. https://doi.org/10.1016/j.ajog.2011.05.032.
[11] Warrick C, Perera L, Murdoch E, Nicholl RM. Guidance for withdrawal and withholding of intensive care as part of neonatal end-of-life care. Br Med Bull 2011;98:99–113. https://doi.org/10.1093/bmb/ldr016.
[12] Gumus II, Uz E, Bavbek N, Kargili A, Yanik B, Turgut FH, et al. Does glomerular hyperfiltration in pregnancy damage the kidney in women with more parities? Int Urol Nephrol 2009;41(4):927–32. https://doi.org/10.1007/s11255-009-9586-1.
[13] Maynard SE, Thadhani R. Pregnancy and the kidney. J Am Soc Nephrol 2009;20(1):14–22. https://doi.org/10.1681/asn.2008050493.
[14] Ross LF. Ethical considerations related to pregnancy in transplant recipients. N Engl J Med 2006; 354(12):1313–6. https://doi.org/10.1056/NEJMsb041648.
[15] Strong C. Minimal risk in research involving pregnant women and fetuses. J Law Med Ethics 2011; 39(3):529–38. https://doi.org/10.1111/j.1748-720X.2011.00619.x.
[16] Piccoli GB, Attini R, Vasario E, Conijn A, Biolcati M, D'Amico F, et al. Pregnancy and chronic kidney disease: a challenge in all CKD stages. Clin J Am Soc Nephrol 2010;5(5):844–55. https://doi.org/10.2215/cjn.07911109.
[17] Hladunewich MA, Bramham K, Jim B, Maynard S. Managing glomerular disease in pregnancy. Nephrol Dial Transplant 2017;32(Suppl. l_1):i48–56. https://doi.org/10.1093/ndt/gfw319.
[18] Piccoli GB, Cabiddu G, Attini R, Vigotti FN, Maxia S, Lepori N, et al. Risk of adverse pregnancy outcomes in women with CKD. J Am Soc Nephrol 2015;26(8):2011–22. https://doi.org/10.1681/asn.2014050459.
[19] Day C, Hewins P, Hildebrand S, Sheikh L, Taylor G, Kilby M, et al. The role of renal biopsy in women with kidney disease identified in pregnancy. Nephrol Dial Transplant 2008;23(1):201–6. https://doi.org/10.1093/ndt/gfm572.
[20] Piccoli GB, Daidola G, Attini R, Parisi S, Fassio F, Naretto C, et al. Kidney biopsy in pregnancy: evidence for counselling? A systematic narrative review. Bjog 2013;120(4):412–27. https://doi.org/10.1111/1471-0528.12111.
[21] Grande JP, Balow JE. Renal biopsy in lupus nephritis. Lupus 1998;7(9):611–7. https://doi.org/10.1191/096120398678920730.
[22] Haładyj E, Cervera R. Do we still need renal biopsy in lupus nephritis? Reumatologia 2016;54(2): 61–6. https://doi.org/10.5114/reum.2016.60214.
[23] Chen TK, Gelber AC, Witter FR, Petri M, Fine DM. Renal biopsy in the management of lupus nephritis during pregnancy. Lupus 2015;24(2):147–54. https://doi.org/10.1177/0961203314551812.
[24] Tedeschi SK, Guan H, Fine A, Costenbader KH, Bermas B. Organ-specific systemic lupus erythematosus activity during pregnancy is associated with adverse pregnancy outcomes. Clin Rheumatol 2016; 35(7):1725–32. https://doi.org/10.1007/s10067-016-3270-5.
[25] Zeisler H, Llurba E, Chantraine F, Vatish M, Staff AC, Sennström M, et al. Predictive value of the sFlt-1:PlGF ratio in women with suspected preeclampsia. N Engl J Med 2016;374(1):13–22. https://doi.org/10.1056/NEJMoa1414838.
[26] Maynard S, Guerrier G, Duffy M. Pregnancy in women with systemic lupus and lupus nephritis. Adv Chronic Kidney Dis 2019;26(5):330–7. https://doi.org/10.1053/j.ackd.2019.08.013.
[27] Lafuente-Ganuza P, Lequerica-Fernandez P, Carretero F, Escudero AI, Martinez-Morillo E, Sabria E, et al. A more accurate prediction to rule in and rule out pre-eclampsia using the sFlt-1/PlGF ratio and NT-proBNP as biomarkers. Clin Chem Lab Med 2020;58(3):399–407. https://doi.org/10.1515/cclm-2019-0939.

[28] Dines V, D'Costa M, Fidler M, Kattah A. The role of kidney biopsy in diagnosis of preeclampsia in kidney transplant patients. Hypertens Pregnancy 2020;39(4):418–22. https://doi.org/10.1080/10641955.2020.1801717.
[29] Fakhouri F, Schwotzer N, Cabiddu G, Barratt J, Legardeur H, Garovic V, et al. Glomerular diseases in pregnancy: pragmatic recommendations for clinical management. Kidney Int 2023;103(2):264–81. https://doi.org/10.1016/j.kint.2022.10.029.
[30] Hahn BH, McMahon MA, Wilkinson A, Wallace WD, Daikh DI, Fitzgerald JD, et al. American College of Rheumatology guidelines for screening, treatment, and management of lupus nephritis. Arthritis Care Res 2012;64(6):797–808. https://doi.org/10.1002/acr.21664.
[31] Hogan JJ, Mocanu M, Berns JS. The native kidney biopsy: update and evidence for best practice. Clin J Am Soc Nephrol 2016;11(2):354–62. https://doi.org/10.2215/cjn.05750515.
[32] Farrington CA. Kidney imaging and biopsy in pregnancy. Adv Chronic Kidney Dis 2020;27(6):525–30. https://doi.org/10.1053/j.ackd.2020.05.012.
[33] Moguel González B, Garcia Nava M, Orozco Guillén OA, Soto Abraham V, Vasquez Jimenez E, Rodriguez Iturbe B, et al. Kidney biopsy during pregnancy: a difficult decision. A case series reporting on 20 patients from Mexico. J Nephrol 2022;35(9):2293–300. https://doi.org/10.1007/s40620-022-01366-1.

CHAPTER 15

Approach to glomerular diseases in pregnancy

Byron F. Sapon Perez
University of San Carlos of Guatemala, Guatemala City, Guatemala

Glomerular diseases in pregnancy

Glomerulonephritis are diseases that affect the glomerulus, altering its permeability, structure, and function. They can be primary, affecting only the kidney, or secondary, forming part of a systemic disease such as systemic lupus erythematosus (SLE) or vasculitis. These glomerulopathies can cause chronic kidney damage, hypertension, proteinuria, and edema, among other symptoms.

Glomerulonephritis is an inflammation of the glomerular tufts (glomeruli). At the level of Bowman's capsule where the ultrafiltrate (urine) is produced, which carries the waste products and the excess fluid that the glomeruli extract from the bloodstream that enters through the afferent capillary system and returns through the efferent capillary system, eliminating it from the body through urine.

In pregnant women, glomerular diseases can have serious consequences for both the mother and the fetus, which can compromise both lives and in these cases the life of the mother prevails, which is why they require early diagnosis and appropriate management.

During pregnancy, glomerulopathies can manifest in various ways. Some are proliferative (with hematuria, hypertension, and decreased glomerular filtration), while others are nonproliferative (with nephrotic syndrome [NS]). These conditions can be diagnosed before or during pregnancy, as well as in the postpartum period. Glomerular diseases encompass a variety of conditions characterized by loss of glomerular architecture or function due to inflammatory processes. Causes include autoimmune diseases, infections, and paraneoplastic conditions.

Diagnosing glomerular diseases during pregnancy can be challenging due to the physiological changes typical of this period and the onset of autoimmune or genetic diseases. Often, clinical symptoms overlap and can be difficult to distinguish. Comprehensive management includes clinical evaluation and histopathological analysis [1].

The glomerular basement membrane, an essential structure in the kidneys, acts as a selective barrier that regulates the passage of molecules through the glomeruli and prevents the passage of larger molecules such as albumin, contributing to adequate kidney function [1]. A small amount of proteins manages to cross the membrane; these proteins

The Kidney of the Critically Ill Pregnant Woman
ISBN 978-0-443-21473-8
https://doi.org/10.1016/B978-0-443-21473-8.00002-1

are partially degraded and reabsorbed by the cells of the proximal tubules, which are the initial segments of the renal tubules. The rest of the proteins are excreted in the urine.

Under normal conditions, the amount of protein in the urine is less than 150 mg/day, measured in a sample collected over 24 h or estimated by the relationship between the concentration of protein and creatinine in a random sample (normal values less than 0.3). Normal albumin excretion is approximately 30 mg/day. When albumin excretion is between 30 and 300 mg/day (20−200 mcg/min), it is called microalbuminuria; if it is greater than 300 mg/day (200 mcg/min), it is called macroalbuminuria.

The presence of protein in the urine, known as proteinuria, can be a sign of various kidney diseases that affect both the glomeruli and the tubules. In some cases, proteinuria occurs along with other urinary disorders, such as hematuria (presence of blood) or glucosuria (presence of glucose). However, there may also be isolated proteinuria, without other obvious symptoms. It is important to consider proteinuria as an indicator of possible kidney damage, especially if it is associated with edema, hypertension, elevated blood creatinine levels, and decreased urine output (oliguria). Glomerular disorders can cause proteinuria, which is often in the nephrotic range ($\geq$3 g/day). The nephrotic range indicates excessive protein loss by the kidney, which can have serious health consequences [2].

Glomerulopathies are classified according to the changes generated in the urine mainly:

- Proteinuria in the nephrotic range and nephrotic urine sediment: These glomerulopathies are called nephrotic and usually cause edema and hypoalbuminemia. Examples include diabetic nephropathy, focal and segmental glomerulosclerosis, and amyloidosis.
- Hematuria: Usually in combination with proteinuria (which may be in the nephrotic range), the red blood cells are generally dysmorphic and there may often be mixed blood or cellular casts (nephritic urinary sediment). These glomerulopathies are called nephritic and usually cause hypertension, elevated serum creatinine, and oliguria. Examples include postinfectious glomerulonephritis, IgA nephropathy, lupus nephritis, and rapidly progressive glomerulonephritis.
- Characteristics of both nephrotic and nephritic types: These glomerulopathies are mixed and have a variable clinical presentation. Some examples are fibrillar and immunotactoid glomerulopathies, membranoproliferative glomerulonephritis (MPGN), and lupus nephritis.

The pathophysiology of nephritic and nephrotic disorders differs markedly, but their clinical overlap is considerable. For example, several disorders may present with the same clinical picture, and the presence of hematuria or proteinuria in itself does not predict response to treatment or prognosis [2]. Therefore, it is necessary to make an etiological diagnosis, that is, identify the cause of glomerulopathy, in order to establish adequate treatment and appropriate follow-up [2].

The causes of glomerulopathies can be primary or secondary. Primary glomerulopathies are those that affect the glomerulus in isolation, without an identifiable systemic cause. Secondary glomerulopathies are those that are the consequence of a systemic disease that affects the glomerulus, such as diabetes, SLE, infection, cancer, hypertension or the use of certain drugs. The causes of glomerulopathies may vary depending on the age and clinical manifestations of the patients.

Although primary glomerulonephritis is not the most common during pregnancy, accurate identification of the diagnosis and its appropriate classification allows optimal management of these conditions. Lupus nephritis has been frequently found through anatomopathological studies, this is explained because SLE tends to affect more frequently women in fertile ages between 10 and 40 years, and is more common in black, indigenous, and Asian women. In the nonpregnant population, the most common primary glomerulopathy is focal and segmental glomerulosclerosis. In Hispanic American women, there is more frequent and severe renal involvement than in other ethnic groups.

Glomerular diseases may be present before or develop during pregnancy. The most common primary glomerulopathies during pregnancy are focal and segmental glomerulosclerosis, membranous glomerulonephritis, IgA nephropathy, MPGN and extracapillary ("crescentic") glomerulonephritis. It has been described that pregnancy in patients with lupus nephritis can lead to more frequent fetal losses, with a greater risk of renal exacerbations, renal deterioration, severe and early preeclampsia (PE), and premature birth, especially when less than 6 months have elapsed of remission, and if pregestational hypertension, antiphospholipid syndrome, anticardiolipin antibodies, or lupus anticoagulant are present. Pregnancy in women with lupus nephritis is successful in a highly variable percentage (20%—90%) according, mainly, to kidney function. Lupus nephritis is not exacerbated by pregnancy [3].

Classification of glomerular diseases

There are different types of glomerular diseases that can affect pregnant women; these can be classified according to their etiology, clinical presentation, and prognosis. Glomerulopathies detected during pregnancy can be both primary and secondary; they can be proliferative (hematuria, hypertension, and decrease in the glomerular filtration index), or nonproliferative; glomerular syndromes such as nephritic syndrome, NS, asymptomatic proteinuria, asymptomatic hematuria; primary or secondary; acute and chronic glomerulonephritis, can occur prior to pregnancy, originating during it or during the postpartum period. For this fact, the importance of evaluating renal parameters is justified both at the beginning of pregnancy, as well as throughout it and in the postpartum period.

Glomerular disorders can be classified based on changes seen in the urine, which reflect the type and degree of glomerular injury. According to this criterion, three types of glomerulopathies can be distinguished:

Nephrotic glomerulopathies: are those that manifest mainly with proteinuria in the nephrotic range and nephrotic urine sediment, which consists of the presence of fatty casts, oval fatty bodies, but few cells or cellular casts. These glomerulopathies often cause NS, which is a set of signs and symptoms that include edema, hypoalbuminemia (low level of albumin in the blood), hypercholesterolemia (high level of cholesterol in the blood), and hypertriglyceridemia (high level of triglycerides in the blood). Some examples of nephrotic glomerulopathies are diabetic nephropathy, focal segmental glomerulosclerosis, and amyloidosis.

Nephritic glomerulopathies: are those that manifest mainly with hematuria, which is the presence of blood in the urine, generally in combination with proteinuria, which may or may not be in the nephrotic range. The red blood cells found in the urine are generally dysmorphic, that is, irregular in shape, and there may often be mixed erythrocytic or cellular casts, which are structures formed by the aggregation of cells and proteins inside the renal tubules. These glomerulopathies often cause nephritic syndrome, which is a set of signs and symptoms that include hypertension, elevated serum creatinine (an indicator of kidney function), and oliguria (decreased urine output). Some examples of nephritic glomerulopathies are postinfectious glomerulonephritis, IgA nephropathy, lupus nephritis, and rapidly progressive glomerulonephritis.

Mixed glomerulopathies: are those that manifest with characteristics of both nephrotic and nephritic syndrome, that is, with proteinuria in the nephrotic range, hematuria, edema, hypoalbuminemia, hypercholesterolemia, hypertriglyceridemia, hypertension, elevated serum creatinine, and oliguria. These glomerulopathies have a variable clinical presentation and an uncertain prognosis. Some examples of mixed glomerulopathies are fibrillar and immunotactoid glomerulopathies, MPGN, and lupus nephritis.

Depending on the time, glomerular disease in pregnancy is classified as acute and chronic:

Acute glomerulonephritis: It is an inflammation of the glomeruli that occurs secondary to an immune reaction to an infection or a systemic disease. It usually manifests with hematuria, proteinuria, hypertension, and decreased kidney function (Nephritic syndrome). It can be complicated by acute kidney injury (AKI), NS, or hemolytic uremic syndrome.

Chronic glomerulonephritis: It is a progressive and irreversible lesion of the glomeruli that leads to the gradual loss of kidney function that can present as a florid condition over a period of at least 3 months. It may be the result of an unresolved range of infections or a primary or secondary disease affecting the kidneys. It is characterized by proteinuria, hypertension, anemia, and electrolyte alterations. It can progress to advanced chronic kidney disease (CKD) and require renal replacement therapy (RRT) such as hemodialysis (HD), peritoneal dialysis (PD), or kidney transplant (RT). Treatment is based on blood pressure (BP) control, kidney protection with inhibitors of the renin-angiotensin-aldosterone system, whenever gestational age or breastfeeding allows, since they are contraindicated because they cause lack of closure of the Ductus arteriosus, and these drugs are excreted by breastfeeding, management of complications, and periodic follow-up; the prognosis depends on the degree and speed of disease progression.

Proteinuria in pregnancy

The mechanisms that can cause proteinuria can be classified into four types:

Glomerular proteinuria: is produced by an increase in the permeability of the glomerular basement membrane, which allows the passage of more proteins into the filtrate. This may be due to different disorders that affect the glomeruli, such as glomerulonephritis, SLE, diabetes mellitus, high BP, etc. Glomerular proteinuria is usually selective, that is, it affects albumin more than other larger proteins. Glomerular proteinuria can be very high, exceeding 3.5 g/day, which is known as NS.

Tubular proteinuria: is caused by an alteration in the reabsorption of proteins in the proximal tubules, which leaves more proteins in the urine. This may be due to different disorders that affect the tubulointerstitial tissue, such as infections, drugs, toxins, hereditary diseases, etc. Tubular proteinuria is usually nonselective, that is, it affects smaller proteins more than albumin. Tubular proteinuria is usually moderate, not exceeding 2 g/day. In addition, tubular proteinuria is often accompanied by other defects in tubular function, such as bicarbonate loss, glycosuria, aminoaciduria, etc.

Overflow proteinuria: is produced by an overload of small proteins in the plasma, which exceed the reabsorption capacity of the proximal tubules. This may be due to excessive production of certain proteins, such as immunoglobulin light chains in multiple myeloma, hemoglobin in hemolysis, myoglobin in rhabdomyolysis, etc. Overflow proteinuria is usually nonselective and moderate, and is associated with a decrease in albumin concentration in the plasma.

Functional proteinuria: occurs due to a transient increase in protein filtration in the glomeruli, due to an increase in renal blood flow. This may be due to physiological situations, such as physical exercise, fever, stress, etc., or pathological situations, such as hyperdynamic heart failure, anemia, hypoxia, etc. Functional proteinuria is usually selective and mild, not exceeding 1 g/day. Functional proteinuria is reversed when renal blood flow is normalized.

Proteinuria can have several negative consequences for health, both at the renal and systemic levels. At the kidney level, proteinuria can cause damage to the glomeruli and tubules, causing progression of kidney disease and a decrease in kidney function. In addition, proteinuria can promote the formation of kidney stones, urinary tract infection, and high BP. At a systemic level, proteinuria can cause a loss of essential proteins for the body, such as albumin, immunoglobulins, hormones, coagulation factors, etc. This can lead to NS, which is characterized by edema, hypoalbuminemia, hyperlipidemia, and thrombosis. It can also cause an alteration of the acid-base balance, a decrease in immunity, an alteration of bone metabolism, anemia, malnutrition, etc.

Glomerular proteinuria appears due to glomerular disorders that typically involve increased glomerular permeability; this permeability allows increased amounts of plasma proteins (sometimes very high amounts) to pass into the filtrate. Tubular proteinuria results from tubulointerstitial disorders that affect protein reabsorption in the proximal tubules, causing proteinuria (mainly from proteins smaller than albumin, such as the light chains of immunoglobulins). The disorders that produce this condition are usually accompanied by other defects in renal function (e.g., bicarbonate loss, glycosuria, amino aciduria) and sometimes by glomerular diseases (which also contribute to proteinuria). Excess flow proteinuria occurs when large amounts of small plasma proteins (e.g., immunoglobulin light chains produced in multiple myeloma) exceed the reabsorption capacity of the proximal tubules.

NS is classically defined by proteinuria >3 g/24 with hypoalbuminemia (serum albumin <3.0 g/dL), edema, hypercholesterolemia, and lipiduria. In normal pregnancy, serum albumin decreases progressively from 38 g/L at week 12–32 g/L at week 36 and cholesterol levels increase from an average of 4.48 mmol/L (173 mg/dL) at week 12, to 6.63 mmol/L (256 mg/dL) at week 36.

The causes of proteinuria in pregnancy, in addition to PE, range from primary glomerular nephropathy (GN) (mesangial IgA GN, focal and segmental GN, minimal change GN, membranous GN, capillary mesangial GN), secondary to rare diseases such as amyloidosis, Fabry disease, Allport syndrome, lupus nephropathy, diabetic nephropathy, HIV nephropathy, nephritis secondary to hepatitis B and C virus infection, or transient causes such as exercise or fever.

It is advisable to determine proteinuria "quantitatively." The "semi-quantitative" option has a high rate of "false positives" and "false negatives." The false positive can appear in the presence of hematuria, urine with a pH > 7, drugs, or urine with a high density (>1.030) [4] and false negatives appear if the urine has a density <1010, or there is hypernatriuria or hyperuricosuria.

Nephrotic syndrome in pregnant women is a rare disease, it is estimated that it occurs in 0.028% of pregnancies and the most frequent cause is PE, although, without histological diagnoses, the incidence of NS secondary to nephropathy. Primary glomerular is difficult to establish, especially if it coexists with PE.

PE is an important cause of fetal morbidity and mortality, due to low birth weight or prematurity; and maternal complications can range from acute renal failure to HELLP syndrome, liver failure, cerebral edema, seizures, or death. In the Western world, it affects 2%–7% of all pregnancies, but in other geographic areas, it can triple. It is a disease that only occurs in the human species, and if it is not treated, it can progress to eclampsia.

NS can occur in patients with PE and also in pregnant patients with primary or secondary glomerulopathies. Fluid retention usually worsens toward the end of pregnancy and in turn aggravates high BP. In patients with suspected glomerular disease detected during pregnancy, kidney biopsy is generally reserved for early stages of pregnancy or for the immediate postpartum. Depending on the underlying disease, corticosteroids or cytotoxic agents may be used during pregnancy (azathioprine).

The causes of proteinuria can be very varied, depending on the mechanism that causes it. The most common causes of proteinuria are those that affect the glomeruli, which usually manifest as an NS. Some of the most common causes of glomerular proteinuria in adults are:

Segmental and focal glomerulosclerosis is a disease that affects some segments of some glomeruli, causing sclerosis or scarring of the glomerular basement membrane. It can be of primary or secondary origin to other diseases, such as obesity, HIV, drug addiction, etc. It manifests itself with massive proteinuria, which can reach the nephrotic range, and a progressive loss of kidney function.

Membranous glomerulonephritis is a disease that affects the glomerular basement membrane, causing thickening and a deposition of immune complexes. It can be of primary or secondary origin to other diseases, such as SLE, hepatitis B, syphilis, cancer, etc. It manifests itself with massive proteinuria, which can reach the nephrotic range, and a variable evolution, which can range from spontaneous remission to terminal renal failure.

Diabetic nephropathy is a complication of diabetes mellitus, which affects the glomeruli and other components of the kidney, causing alteration of kidney structure and function. It is characterized by the presence of nodules or deposits of hyaline material in the glomerular basement membrane, which increase its permeability. It manifests itself with proteinuria, which can reach the nephrotic range, and a progressive loss of kidney function, which can lead to end-stage renal failure.

Renal amyloidosis is a disease caused by the accumulation of an abnormal substance called amyloid in the glomeruli and other organs.

Diagnostic approach

Glomerulonephritis are a heterogeneous group of diseases that requires a careful diagnostic approach and precise classification to guide appropriate treatment. Glomerulonephritis can be classified into several clinical patterns, including asymptomatic urinary disorders, macroscopic hematuria, NS, nephritic syndrome, rapidly progressive glomerulonephritis, and chronic glomerulonephritis.

Histological findings are essential for diagnosis and treatment, providing information on the presence of acute or chronic lesions and the degree of kidney tissue involvement. Immunofluorescence (IF) and electron microscopy (EM) techniques allow the localization of immune complex deposits in different regions of the glomerulus, allowing a new diagnostic approach.

According to its clinical evolution, glomerulonephritis can be classified as acute, subacute, and chronic. Although this classification is useful, in clinical practice, these presentations can overlap and be modified by rapidly progressive behavior at some point in their evolution.

There is no absolute indication for renal biopsy during pregnancy. However, it must be performed to distinguish and classify severe nephropathies. When there is rapid deterioration of kidney function, termination of pregnancy may be required. Renal biopsy is a procedure with morbidity, which should be considered only if it offers the patient the opportunity for better management. In NS, it is necessary to weigh the need for rapid aggressive treatment or wait for the termination of the pregnancy, to perform the biopsy and start treatment (usually immunosuppressive). The usefulness of renal biopsy, specifically IF, to clarify the diagnosis is highlighted. Kidney biopsy can provide valuable information about the nature and severity of glomerular disease, which can help guide treatment.

The diagnosis of glomerulopathy should be based on clinical history, physical examination, laboratory tests, and renal biopsy when possible and safe [2]. Kidney biopsy can help determine the type, severity, and activity of glomerulopathy, as well as guide treatment and prognosis. However, kidney biopsy has risks of complications, especially in pregnancy, such as bleeding, infection, rupture of membranes, and fetal loss. Therefore, kidney biopsy should be performed only when the benefits outweigh the risks, and with the patient's informed consent.

Historically, glomerulonephritis has been classified according to its histological pattern, using light microscope, IF, and EM techniques. However, this classification may be limited, since a single pattern does not always correspond to a single cause or clinical presentation.

Recently, a new classification based on the pathogenic mechanisms of glomerulonephritis has been proposed. For example, MPGN is now classified according to the presence of polyclonal or monoclonal immunoglobulins (Ig) and/or C3 deposition on IF studies. In addition, a new clinical entity, C3 glomerulopathy, has been identified, which is believed to be caused by alteration of the alternative complement pathway.

Age is also a determining factor in the evaluation of the glomerular process. For example, NS in children is usually secondary to minimal change (MC) glomerulonephritis, while in adults, membranous nephropathy and diabetic nephropathy are the main causes. Glomerulonephritis can be classified prognostically based on the magnitude of mean proteinuria throughout follow-up, the glomerular process diagnosed, and the sex of the patient.

Women with CKD should not become pregnant. If this is the case, it will always be considered a high obstetric risk pregnancy. Maternal and fetal prognosis worsens according to the severity of CKD. Obstetric complications occur in about half of patients with moderate or severe chronic kidney failure, including HELLP syndrome. There is only one glomerular disease exclusive to pregnancy and it is "pres-eclamptic glomerulopathies" (usually given the imprecise term "endotheliosis") [1,5]. Many studies have evaluated whether pregnancy adversely affects the natural course of glomerular diseases and whether the course of pregnancy is influenced by glomerulopathy; however, there are few controlled studies that allow clear conclusions to be drawn.

Experimental data suggest adverse effects of "abnormal" pregnancy hemodynamics. The risk of developing CKD was not increased in 148 pregnant women with various biopsy-proven kidney diseases, compared with a control group of women with similar diseases, but not pregnant. There is usually a transient worsening of hypertension, proteinuria, and renal dysfunction in patients with preexisting glomerulopathies [2,3,5–7]. Patients with good glomerular filtration before conception, there are usually no effects of pregnancy on kidney disease. Uncontrolled hypertension, proteinuria in the nephrotic range, or impaired renal function at the time of conception or in the early stages of pregnancy are associated with an increased risk of deterioration in function.

It is important that pregnant women with glomerulopathies receive specialized medical care and regular monitoring of their kidney function, nutritional status, BP, and fetal well-being. It is also recommended that they receive advice on the risks and benefits of continuing or terminating the pregnancy, as well as on family planning and contraception [2].

Therapeutic approach

Glomerulopathies can complicate pregnancy or be caused by pregnancy. Some of the complications that can occur are PE, eclampsia, HELLP syndrome, spontaneous abortion, intrauterine growth retardation, fetal death, acute or chronic renal failure, and maternal death [2,5]. The management of glomerulopathies in pregnancy depends on the type, severity, etiology, and timing of the disease. Some of the general measures that can be taken are BP control, the use of corticosteroids or immunosuppressants, induction of labor, dialysis, or KT [2,3].

In patients with membranous GN, IgA nephropathy, and proliferative GN, the incidence of normal delivery is between 71% and 80%. In GEFyS, there is a higher incidence of maternal and fetal complications, with a worse prognosis than membranous GN and IgA nephropathy. In granulomatous polyangiitis and extracapillary GN, increased risk of renal disease relapses and increased risk of complications and fetal deaths are also reported. In addition to fetal loss, premature birth and intrauterine growth retardation are reported as complications. Approximately 10%–15% fetal losses are reported in the GN group.

The management of patients who are critically ill due to glomerulopathies during pregnancy is a clinical challenge that requires multidisciplinary care. Glomerulopathies can complicate pregnancy with risks for both the mother and the fetus, such as PE, intrauterine growth restriction, premature birth, acute and chronic renal failure, thrombosis, and hemorrhage [3,6,7].

Treatment of glomerulopathy should be individualized based on the type, severity, and activity of the disease, as well as the patient's condition and gestational age. The goal of treatment is to control symptoms, prevent or delay the progression of kidney disease, and improve maternal and fetal outcomes. Treatment may include nonpharmacological measures, such as salt and fluid restriction, weight control, low protein diet always with nutrition intervention for weight monitoring, and iron and folic acid supplements [2,6,7].

Pharmacological treatment should take into account the safety and effectiveness of medications for both the mother and the fetus, and avoid those that may be teratogenic, nephrotoxic, or hypotensive. Some of the medications that can be used are antihypertensives, diuretics, corticosteroids, immunosuppressants, and anticoagulants, as the case warrants. Pharmacological treatment must be adjusted periodically according to clinical evolution and laboratory tests [2,5,7,8].

Supportive care includes frequent monitoring of BP, weight, edema, proteinuria, renal function, hemoglobin, platelet count, clotting factors, lipid profile, complement, and liver function. It also includes fetal monitoring with Doppler ultrasound, cardiotocography, and biophysical profile, to evaluate fetal growth, vitality, and well-being. Supportive care should be provided by a multidisciplinary team that includes nephrologists, obstetricians, neonatologists, nurses, and psychologists, who can offer comprehensive and coordinated care to the patient and her family [2,3,5].

The timing and route of delivery should be determined according to the clinical status of the mother and fetus, and the type and severity of glomerulopathy [1–3,6,7]. The birth must be planned in advance and in a specialized center, with availability of maternal and neonatal intensive care [1,2]. Delivery is preferably by cesarean section, whenever possible and safe, to avoid surgical and anesthetic complications [2]. Labor should be induced or brought forward if there are signs of maternal or fetal deterioration, such as severe PE, acute renal failure, thrombosis, hemorrhage, intrauterine growth restriction, fetal distress, or death [1,2,7,8].

The management of patients with glomerulopathies after delivery should include evaluation of renal function, proteinuria, BP, anemia, coagulation, and immunity [1–3]. Pharmacological treatment should be reviewed and modified according to need and breastfeeding [6]. Breastfeeding is beneficial for both mother and baby, but should be avoided if the mother is receiving medications that can pass into breast milk and cause adverse effects on the baby. Long-term follow-up should be carried out by the nephrologist, with periodic controls of kidney function and cardiovascular risk [1,2,6,7].

It is important to remember that each patient is unique and treatment must be personalized based on the patient's individual needs. Collaboration between nephrologists and obstetricians is essential for the optimal management of these patients. Along with the specific treatment of each glomerulonephritis, symptomatic therapy should be instituted depending on the clinical manifestations. Treatment of glomerular disease during pregnancy can be challenging due to the need to balance the risks and benefits for the mother and fetus. Here are some general recommendations based on the medical literature [3,6,9]:

The treatment of primary glomerulonephritis includes general measures such as diet and obesity control, smoking cessation, and regular physical exercise. Treatment of hypertension, control of proteinuria, treatment of hyperlipidemia, and treatment of edema may also be required. The diet may involve restriction of salt and fluids, as well as protein intake. The drugs of choice for the treatment of hypertension are ACE inhibitors and angiotensin II receptor antagonists. The objective of using these medications is to reduce proteinuria, in addition to controlling BP. Hyperlipidemia, a common disorder in glomerulonephritis that presents with NS, can be treated with HMGCoA reductase inhibitors (statins). In case of significant edema, loop diuretics can be used.

In the presence of nephritic syndrome, a moderate restriction of salt (2–4 g/day) and fluids will be necessary; only in cases of significant edema are more severe restrictions applied. Protein intake, in NS, a diet of 0.8 g/kg/day is recommended. If there is renal failure, reduce intake to 60 g/day.

BP must be strictly controlled. Angiotensin-converting enzyme inhibitors (ACE inhibitors) and angiotensin II receptor blockers (ARBs) are widely recommended for NS due to their nephroprotective effects. However, these medications are not safe during pregnancy and should be stopped before conception or as soon as pregnancy is confirmed. The objective of using ACEI/ARB is to reduce proteinuria, in addition to controlling BP.

In NS with significant edema, loop diuretics can be used. In NS with significant edema, initial loop diuretics at higher doses (furosemide or torsemide) can be used. In case of anasarca or albumin <2 g/dL, the combined infusion of albumin and intravenous furosemide may be useful. If the response to loop diuretics is poor, thiazides or antialdosterone drugs may be added.

Thromboembolic complications: Renal vein thrombosis and thromboembolic phenomena in general constitute one of the most serious complications of NS. The frequency ranges between 5% and 60%. The risk increases with severe hypoalbuminemia (<2 g/dL). Antithrombotic prophylaxis (low molecular weight heparin) is sometimes necessary, especially in patients who require rest. In the event of any thrombotic episode, anticoagulant treatment should be initiated. Anticoagulants for membranous glomerulonephritis with high risk of renal or systemic venous thrombosis should be evaluated individually, as they may increase the risk of maternal and fetal bleeding.

The basic point in the treatment of nephritic syndrome is the control of edema, monitoring the possible appearance of heart failure and hypertension. Once acute renal failure is established, treatment consists of a series of medical measures and dialysis support, if necessary, until the condition resolves. In glomerulonephritis that causes chronic kidney failure, treatment seeks to prevent the appearance of symptoms derived from this condition, minimize complications, and preserve kidney function.

In gross hematuria, the outbreaks are self-limiting and do not require treatment; however, it is advisable to ensure abundant fluid intake and alkalinize the urine in severe cases. There is also no specific treatment in case of isolated hematuria.

Treatment of glomerulopathies in pregnancy depends on the type, cause, and severity of kidney disease, as well as maternal and fetal conditions. In general, multidisciplinary management is recommended that includes nephrologists, obstetricians, neonatologists, and pathologists. Some general measures are [9]:

- Avoid nonsteroidal antiinflammatory drugs (NSAIDs), which can worsen kidney function and cause adverse effects on the fetus.
- Correct hydroelectrolyte and acid-base alterations, as well as anemia and malnutrition.
- Prevent and treat urinary tract infections, which can aggravate kidney damage and increase the risk of premature birth.
- Monitor fetal growth and well-being with periodic ultrasounds and fetal vitality tests.
- Plan the time and route of delivery according to clinical and obstetric conditions.
- Omega-3 polyunsaturated fatty acids (fish oil) for IgA nephropathy may reduce proteinuria and disease progression.

In some cases, specific treatment may be indicated depending on the type of glomerulopathy, such as: immunosuppressants (corticosteroids, mycophenolate, cyclosporine, etc.) for glomerulopathies of autoimmune origin, such as lupus nephritis, rapidly progressive glomerulonephritis, or C3 glomerulopathy. These drugs should be used with caution and under strict medical supervision, as they may have side effects and teratogenicity.

Regarding immunosuppressive treatment, pulsed methylprednisolone and plasmapheresis are options that can be applied for a faster effect. Regarding treatment, most immunosuppressants are teratogenic, but we have azathioprine and cyclosporine as the most used drugs, since their safety during pregnancy has been demonstrated. Hydroxychloroquine currently represents the first therapeutic option for SLE during pregnancy. It is safe for pregnancy and has been shown to be associated with a reduced risk of congenital heart block in newborns.

Azathioprine is frequently the drug of choice to maintain the disease at rest during pregnancy, and therefore it is one of the drugs of choice in maintenance therapy. Regarding exposure to cyclophosphamide in the last trimester, it has been related to intrauterine growth restriction, suppression of hematopoiesis, and neurological deterioration. According to the EULAR/ACR guideline [10] and the results of the Euro-Lupus Nephritis Trial (ELNT) [11], it is the therapy of choice in the induction of remission compared to the combination of azathioprine with tacrolimus.

Another immunosuppressive drug to consider for maintenance therapy is tacrolimus, given its safety profile during pregnancy. Current evidence indicates that with tacrolimus there is no increase in the rate of congenital malformations [12]. Worsening proteinuria in pregnancy poses a significant diagnostic and therapeutic challenge. Although lupus nephritis and vasculitis can debut during pregnancy, the effect of this on other primary glomerular diseases is unclear. Currently there are no other biomarkers established for use in pregnancy. Although M-type antiphospholipase A2 receptor has been described in membranous nephropathy [13] and soluble urokinase plasminogen activator receptor in focal and segmental glomerulosclerosis, their use in pregnancy requires further study.

Continuing with the treatment, the use of diuretics from the sulfonamide group, such as furosemide, has shown that they are not teratogenic, but their use should be limited to severe edema and management on a case-by-case basis, taking care not to produce hypovolemia and placental hypoperfusion, with daily hydroelectrolyte balance. Regarding neonatal complications, women with antiphospholipid syndrome present a risk of developing fetal heart block due to placental transfer of immunoglobulins that leads to endocardial fibroelastosis, estimated at 5% of cases, although it has only been observed with titers $\geq$50 U/mL. Therefore, treatment with hydroxychloroquine is recommended in pregnant women with SLE.

Some authors demonstrated that thiazides decreased placental perfusion, and it has been suggested that premature infants of mothers receiving hydrochlorothiazides would be at greater risk of developing hyperbilirubinemia than with furosemide. Among the maternal adverse effects are hypokalemia, hypomagnesemia, hyponatremia, hyperuricemia, and hypovolemia, while among its neonatal adverse effects are: hypoglycemia, thrombocytopenia, hydroelectrolyte imbalances, jaundice, and small babies for gestational age. On the other hand, in relation to loop diuretics (furosemide) during pregnancy, it is limited to hypertensive pregnant women who present pulmonary edema, heart disease, massive fluid retention, or kidney injury with hydrosaline retention and play an important role in the management of postpartum hypertension.

Some authors warn that diuretics can complicate the course of anesthesia. This fact must be taken into account, especially when the route of delivery is not completely decided in advance. The usual oral dose of hydrochlorothiazide is 25 mg/day to a maximum of 150 mg/day. The commonly used dose of furosemide is 40 mg up to a maximum of 120 mg/day. The goal of treatment for the patient with severe NS is to maintain the level of edema at a tolerable level by losing 2–3 kg, without seeking to eliminate all edema. It may be necessary to administer albumin in patients with severe hypoalbuminemia.

There is no experience with the use of biological drugs in pregnant women. Mycophenolate mofetil as an antiproliferative medication is contraindicated and even women with lupus glomerulonephritis who have pregnancy plans are insisted that they must stick to one of the many methods of contraception due to the teratogenic nature of mycophenolate mofetil, otherwise they must give advance notice at least 6 months in advance if you have pregnancy plans, since the immunosuppressive regimen must be changed, especially mycophenolate mofetil for azathioprine, and folic acid must be added. It is believed that women with MPGN, focal and segmental sclerosis, and reflux nephropathy would have a greater risk of deteriorating their kidney function after pregnancy, whether or not they had AKI.

In primary glomerulonephritis, the use of immunosuppressants has limitations: mycophenolic acid and cyclophosphamide are teratogens and anti-CD20 type monoclonal antibodies (Rituximab) cross the fetoplacental barrier and can induce transient depletion of

B cells in the fetus. Calcineurin inhibitors (Cyclosporin A) can be used; there is no evidence of teratogenicity since the P-glycoprotein inactivates it as it passes through the placenta, short half-life steroids (Prednisone and methylprednisolone) can also be used due to its inactivation by placental hydroxysteroid dehydrogenase type 2, which prevents the adverse effects of steroids on the fetus. Calcineurin inhibitors (Cyclosporin A) can be used; there is no evidence of teratogenicity since P-glycoprotein inactivates it as it passes through the placenta.

Nephrotic syndrome in pregnancy

The basic pathogenic phenomenon of PE is endothelial dysfunction (ED) conditioned by an "imbalance" of angiogenic factors that produce placentation anomalies and placental ischemia, development of a "maternal syndrome," and alterations in immunological tolerance at the uteroplacental interface. It is not a limited process that is resolved with childbirth, since ED can persist and become a future cardiovascular risk factor.

The differential diagnosis of NS with PE is based on clinical manifestations. It usually appears after the 20th week of gestation with proteinuria between 2 and 5 times the basal proteinuria, with arterial hypertension, severe headache, epigastric pain, and usually with thrombocytopenia and elevated transaminases. The therapeutic objective in PE should be primary prevention. Adequate prenatal care with special monitoring of weight gain, and monitoring of blood pression and control of proteinuria, can reduce the incidence and severity of PE.

Some early diagnosis methods have been studied, among which the following can be mentioned: elevated sFlt1 (fms-type tyrosine kinase1), elevated sFlt1/PIFG (placental growth factor) index, decreased PAPP-A (plasma metalloproteinase associated with protein A in pregnancy), VEGF (vascular endothelial growth factor) polymorphism, angiotensin II polymorphism, decreased eNOS (nitric oxide synthases), reduced MTHFR (methylenetetrahydrofolate reductase), increased proinflammatory cytokines and elevated plasma ADMA (asymmetric dimethyl arginine) concentration.

None of the test has the sensitivity and specificity necessary for use in daily clinical practice [6] although with the combination of several and UTPi (Uterine Artery Pulsatility index), the detection rate can reach 74.8%. Other markers are the determination of Serpin 1 [2], which is detectable 10 weeks before the appearance of PE, also podocyturia with a sensitivity of 88% and specificity of 79% and the Hyperbaric Index (HIB), this is a new concept, derived from ABPM (Ambulatory Blood Pressure Monitoring) defined as "area of excess blood pressure above the upper limit of the tolerance interval."

In the normal pregnant woman, the BP profile is a decrease in SBP/DBP until week 20, followed by an increase in it until delivery. The "predictable variability" of BP throughout pregnancy in normotensive pregnant women is an increase of 7% in SBP and 9% for DBP. The pattern of BP in women with PE or gestational hypertension is different, but also predictable. The SBP in the first half of pregnancy is slightly higher in PE than in women with gestational hypertension and in the second half the elevation is also higher in those who develop PE than in gestational hypertension. BP differences between normotensive pregnant women and complicated pregnant women (PE or gestational hypertension) are detected from the first trimester with significant differences in the circadian cycle of 12 mmHg in SBP and 7 mmHg in DBP. With these "patterns" of BP in normal pregnant women, such as those with hypertension or those who develop PE, ABPM can now be used as a predictor of PE.

PE nephropathy is widely considered a clear example of "endothelial disease" or "endotheliosis" understood as the "swelling" of endothelial cells due to cytoplasmic expansion, proliferation of mesangial cells without thickening of the basement membrane, and fibrinogen deposit. Endotheliosis exclusively expresses endothelial cell injury and the activation of intravascular coagulation, but there is no evidence of immunological injury.

The basement membrane and podocytes are not affected as usually observed in EM images. However, despite this histological demonstration, there is evidence that podocyte alteration is basic in the development of PE nephropathy. In different experiences, a low expression of nephrin and synaptopodin was observed, which could be the cause of proteinuria. The loss of nephrin could be due to a blockade of circulating VEGF by sFlt-1 and thus produce greater expression of endothelin −1, leading to podocyte injury.

In PE, in addition to the typical nephropathy or "endotheliosis," focal and segmental glomerulonephritis (FSGS) may develop. This could be a secondary manifestation of severe systemic hypertension and hyperfiltration, and is self-limiting, but proteinuria may persist up to 30 months postpartum. The differential diagnosis versus "de Novo" FSGS is not easy, but the coexistence of global sclerosis-type histological lesions and tubular/interstitial atrophy indicates that it is presumably a primary glomerular lesion.

In FSGS, there is a high incidence of spontaneous abortion (5.9%—28.6%), premature birth in 5.9%—58.8% of cases, and perinatal death between 11.7% and 45.2%. Gestational hypertension appears in 20%—71% of pregnant women and persists for more than 6 months in 4%—66% of women. In MC glomerulonephritis (MC), a patient increases proteinuria in the postpartum and perinatal death occurred in 14.3% of cases. In Membranous Nephropathy (MN) [1—15], the low incidence of abortions and perinatal death is notable. Gestational hypertension is recorded in 7.7%—45.4% of pregnancies. In capillary mesangioglomerulonephritis (MPGN), premature birth occurs in 16.7%—23.8% and fetal death is less frequent than in FSGS and MN, ranging between 0% and 11%.

The treatment of NS in pregnancy has some particularities. The use of diuretics should be reserved for NS refractory to immunosuppressants and with strict control over maternal-fetal hydroelectrolyte disorders. The use of parenteral albumin is not indicated and may also increase hyperfiltration. When plasma albumin is less than 2 gr/L, the hypercoagulable state should be treated with low molecular weight heparin. As primary prevention of PE in high-risk pregnant women (HIB >12) from the 11th week of gestation, acetylsalicylic acid should be used at a dose of 100 mg/24 h/night.

The appearance of NS due to primary GN is a rare entity in pregnancy, although it can present with superimposed PE which makes diagnosis difficult. The incidence of PE ranges between 2% and 7% of pregnancies, and primary prevention is very important not only because of the maternal-fetal complications it produces but also because it is a cardiovascular risk factor that persists beyond pregnancy.

Currently, the early diagnostic method that is easy to apply clinically is the determination of the IHB, which makes preventive treatment with acetyl salicylic acid possible. In the case of NS due to primary GN, minimal change glomerulonephritis is the most favorable and those with the worst prognosis are mesangiocapillary GN and focal and segmental GN. In relation to treatment, the use of diuretics, parenteral albumin, and immunosuppressants except calcineurin inhibitors should be avoided and the consequences of hypercoagulability should be prevented. Similarly, calcium supplementation has been shown to reduce the incidence of PE more pronouncedly in those women with low calcium intake.

Nephritic syndrome in pregnancy

SLE is a multisystem autoimmune disease that primarily affects women of childbearing age. Lupus nephritis is the most common clinical manifestation during pregnancy and constitutes a risk factor for pregnancy loss, especially in patients with renal failure.

In addition, it presents an increased risk of fetal loss, intrauterine growth restriction, and hypertension. Women with SLE have a normal fertility rate, making pregnancy a topic of frequent interest in these patients. According to the different sources consulted, SLE has an estimated prevalence of 20–150 cases per 100,000 and an incidence of 1–25 cases per 100,000 people/year, and 1/1000 women between 15 and 45 years old.

The incidence of SLE is estimated at 9 per 10,000 inhabitants, mainly affecting women of childbearing age, with a female/male ratio of 10/1, which reflects a relationship between sexual hormones and the etiology and pathophysiology of autoimmunity. Pregnancy and SLE are affected reciprocally. Pregnancy can increase the activity of SLE and have short- or long-term adverse effects on kidney function. Lupus nephritis is the third form of clinical presentation of SLE outside of pregnancy, but during pregnancy it represents the most frequent clinical manifestation. Lupus nephritis is a risk factor for pregnancy loss, especially in patients with renal failure. In addition, it presents an increased risk of fetal loss, restricted intrauterine growth (CIR) and gestational hypertension, in 30%–50%, even in patients with lupus nephritis with nonactive SLE.

The best time for conception is considered to be after 6 months of clinical stability, since adverse pregnancy outcomes are related to active nephritis. This may be due to compromised microcirculation in the uteroplacental vessels of SLE patients with arterial hypertension and lupus nephritis.

The diagnosis of lupus nephritis in a healthy pregnant woman is a significant challenge, since it is usually considered a second- or third-line diagnosis, once other more common pathologies have been ruled out. Interdisciplinary management helps establish an earlier diagnosis and facilitates early therapeutic initiation. It seems that the role of complement in the development of the placenta and the normal fetus is important. SLE flares are often associated with low levels of C3 and C4, and normal C3 and C4 values are associated with reduced SLE flares.

Although C4 and C3 levels can be affected by pregnancy itself. Rates of renal exacerbation and de novo disease were significantly higher in those with positive anti-DNA and low complement or low complement alone, than in patients without serological activity at the beginning of the study. The differential diagnosis of the onset of lupus nephritis during pregnancy continues to be a challenge, despite advances in angiogenic markers. Clinical assessment constitutes the cornerstone of this diagnostic process and its implications in terms of current and future pregnancy complications [14,15].

References

[1] Kidney Disease: Improving Global Outcomes (KDIGO) Glomerular Diseases Work Group. KDIGO 2021 clinical practice guideline for the management of glomerular diseases. Kidney Int October 2021; 100(4S):S1–276. https://doi.org/10.1016/j.kint.2021.05.021. PMID: 34556256.

[2] Proteinuria MG. Urogenital disorders. MSD Manual Professional Edition Saint Louis University Revised/Revised; December 2022.

[3] Martínez P, Rodríguez D. Glomerulopathies: concept, classification, etiopathogenesis. (Glomerulopatías. Concepto. Clasificación. Etiopatogenia). Medicine: Programa de Formación Médica Continuada Acreditado 2007;9(80):5125–30. https://doi.org/10.1016/S0211-3449(07)74622-2. ISSN 0304-5412.

[4] Pereira MM, Torrado J, Sosa C, Zócalo Y, Bia D. Shedding light on the pathophysiology of preeclampsia-syndrome in the era of cardio-obstetrics: role of inflammation and endothelial dysfunction. Curr Hypertens Rev 2022;18(1):17–33. https://doi.org/10.2174/15734021 17666210218105951. PMID: 33602097.

[5] Lau CS, Yin G, Mok MY. Ethnic and geographical differences in systemic lupus erythematosus: an overview. Lupus 2006;15(11):715–9. https://doi.org/10.1177/0961203306072311. PMID: 17153840.

[6] Bernasconi DA. Kidney and pregnancy. Glomerulopathies and pregnancy. Hospital juan A. Fernández autonomous city of Argentina. Electron J Biomed 2016;3:27–32. ISSN: 1697-090X. Disponible en: https://biomed.uninet.edu/2016/n3/2bernasconi.html.

[7] Mejía G, Builes M, Arbelaez M, Henao JE, Arango JL, García A. Clinicopathological description of glomerular diseases, Study of 383 kidney biopsies. (Descripción clínico-patológica de las enfermedades glomerulares. Estudio de 383 biopsias renales). Acta Méd Colomb 1989;14:369–74.

[8] Gómez-Jiménez JM, Arias LF. Glomerular diseases during pregnancy. Case series and literature review. Colomb J Obstet Gynecol 2008;59(4):343–8. ISSN: 0034-7434. Disponible en: https://www.redalyc.org/articulo.oa?id=195214333011.

[9] Otero GA. Sindrome nefrotico y embarazo (Nephrotic syndrome and pregnancy). Hipertens Riesgo Vasc 2023;40(1):1–4. https://doi.org/10.1016/j.hipert.2022.12.001. Spanish. PMID: 36906347.

[10] Barbhaiya M, Zuily S, Naden R, Hendry A, Manneville F, Amigo MC, Amoura Z, Andrade D, Andreoli L, Artim-Esen B, Atsumi T, Avcin T, Belmont HM, Bertolaccini ML, Branch DW, Carvalheiras G, Casini A, Cervera R, Cohen H, Costedoat-Chalumeau N, Crowther M, de Jesús G, Delluc A, Desai S, Sancho M, Devreese KM, Diz-Kucukkaya R, Duarte-García A, Frances C, Garcia D, Gris JC, Jordan N, Leaf RK, Kello N, Knight JS, Laskin C, Lee AI, Legault K, Levine SR, Levy RA, Limper M, Lockshin MD, Mayer-Pickel K, Musial J, Meroni PL, Orsolini G, Ortel TL, Pengo V, Petri M, Pons-Estel G, Gomez-Puerta JA, Raimboug Q, Roubey R, Sanna G, Seshan SV, Sciascia S, Tektonidou MG, Tincani A, Wahl D, Willis R, Yelnik C, Zuily C, Guillemin F, Costenbader K, Erkan D, ACR/EULAR APS Classification Criteria Collaborators. 2023 ACR/EULAR antiphospholipid syndrome classification criteria. Ann Rheum Dis October 2023;82(10):1258−70. https://doi.org/10.1136/ard-2023-224609. PMID: 37640450.

[11] Houssiau FA, Vasconcelos C, D'Cruz D, Sebastiani GD, Garrido Ed Ede R, Danieli MG, Abramovicz D, Blockmans D, Mathieu A, Direskeneli H, Galeazzi M, Gül A, Levy Y, Petera P, Popovic R, Petrovic R, Sinico RA, Cattaneo R, Font J, Depresseux G, Cosyns JP, Cervera R. Immunosuppressive therapy in lupus nephritis: the Euro-Lupus Nephritis Trial, a randomized trial of low-dose versus high-dose intravenous cyclophosphamide. Arthritis Rheum August 2002;46(8):2121−31. https://doi.org/10.1002/art.10461. PMID: 12209517.

[12] Ronco P, Debiec H. Pathophysiological advances in membranous nephropathy: it is time for a change in patient care. Lancet May 16, 2015;385(9981):1983−92. https://doi.org/10.1016/S0140-6736(15)607310.

[13] Patel M, Clarke AM, Bruce IN, Symmons DP. The prevalence and incidence of biopsy-proven lupus nephritis in the UK: evidence of an ethnic gradient. Arthritis Rheum September 2006;54(9):2963−9. https://doi.org/10.1002/art.22079. PMID: 16947632.

[14] Hannah J, Casian A, D'Cruz D. Tacrolimus use in lupus nephritis: a systematic review and meta-analysis. Autoimmun Rev January 2016;15(1):93−101. https://doi.org/10.1016/j.autrev.2015.09.006. Epub 2015 Sep 30. PMID: 26427983.

[15] Cavalli S, Lonati PA, Gerosa M, Caporali R, Cimaz R, Chighizola CB. Beyond systemic lupus erythematosus and anti-phospholipid syndrome: the relevance of complement from pathogenesis to pregnancy outcome in other systemic rheumatologic diseases. Front Pharmacol February 15, 2022;13:841785. https://doi.org/10.3389/fphar.2022.841785. PMID: 35242041; PMCID: PMC8886148.

CHAPTER 16

Continuous renal replacement therapy and pregnancy

Ranajit Chatterjee[1] and Priyanka H. Chhabra[2]
[1]Swami Dayanand Hospital, New Delhi, Delhi, India; [2]Department of Anaesthesia, VMMC-Safdarjung Hospital, Delhi, India

Introduction

Renal replacement therapy is required for optimal management of renal dysfunction in pregnancy. Structural and/or functional loss of kidney function leads to either rise in serum creatinine or decrease in urine output or both. There is also a progressive decrease in glomerular filtration rate (GFR). Diabetes, hypertension, and obesity are commonly associated with kidney disease. Often, renal dysfunction poses risks to both the mother and the fetus. Several complications are linked to renal dysfunction like abruptio-placentae, puerperal sepsis, septic abortion, HELLP syndrome, etc., and are associated with worse outcomes in the fetus. An important cause of acute kidney injury (AKI) during pregnancy is pregnancy-associated thrombotic microangiopathies (P-TMAs), which includes hemolytic uremic syndrome (HUS) and thrombotic thrombocytopenic purpura (TTP). Also, urinary tract infections (UTIs) during pregnancy also pose a serious risk. Appropriate antibiotics are cornerstone in management of UTIs.

Physiological changes in kidneys during pregnancy

Pregnancy is associated with a multitude of physiological changes during pregnancy. The anatomical size of kidneys is increased by 1−1.5 cm. Ureters and renal pelvis undergo dilatation under the effect of progesterone, more so on right side due to dextroversion of uterus. Left side ureters are largely spared due to sigmoid colon. Fig. 16.1 shows changes in renal vasculature during pregnancy. These changes promote urinary stasis in pelvicalyceal system leading to UTIs and renal stones. GFR increases by 50% and renal plasma flow (RPF) increases by up to 85% above prepregnant level. This is the consequence of an increase in the overall blood volume and renal vasodilatation with marked reduction in renal afferent and efferent arteriolar resistance. Renal hyperperfusion determines the renal hyperfiltration in pregnancy. This leads to an increased urine output and urinary frequency in pregnant woman. As a consequence, creatinine clearance increases to 150−200 mL/min. Serum creatinine falls by 0.3−0.4 mg/dL in pregnancy. Hence, serum creatinine levels of 1.0 mg/dL which seems otherwise innocuous in nonpregnant individuals might depict the onset of renal dysfunction in pregnancy [1]. A proteinuria of

The Kidney of the Critically Ill Pregnant Woman
ISBN 978-0-443-21473-8
https://doi.org/10.1016/B978-0-443-21473-8.00001-X

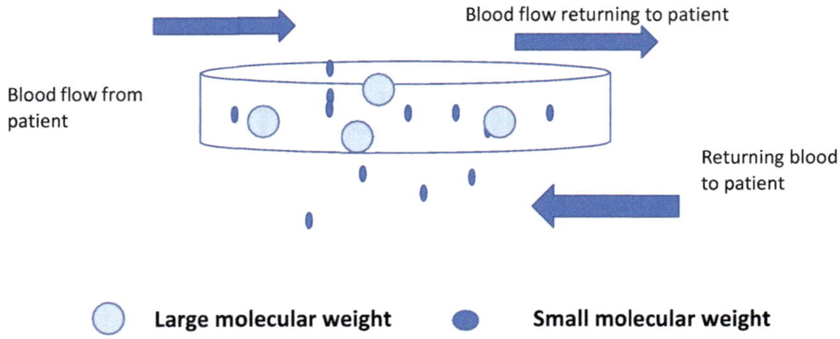

Figure 16.1 Diffusion—removal of small molecular weight solutes across semipermeable membrane.

Table 16.1 Various anatomical and physiological changes during pregnancy.

Anatomical changes	Physiological changes
Increased size of kidneys Dilatation of ureters, renal pelvicalyceal System, hydronephrosis	GFR increases by 50% RPF increases by 85% Cr falls by 0.3–0.5 mg/dL Sodium falls by 4–5 mosm/L Osmolality also falls around 270 mosm/L

300 mg/dL is considered normal in pregnancy. Table 16.1 depicts the physiological changes associated with pregnancy. The physiological changes of pregnancy are further elucidated in Table 16.1.

Epidemiology

Renal dysfunction in pregnancy has a bimodal onset with one peak occurring in first trimester due to septic abortions, lupus nephritis, and hyperemesis gravidarum. The second one is seen in last trimester owing to preeclampsia, HUS, TTP, acute fatty liver of pregnancy (AFLP), and several other obstetric complications. Its incidence in India is estimated to be less than 1% [2]. The incidence of PR-AKI has declined from 1 in 3000 population in the past to 1 in 20,000 in recent years owing to better understanding, recognition, prevention, and improved management options. However, it still continues to be a major public health problem.

Diagnostic definition of renal diseases in pregnancy is not uniform and validated. General consensus is to define pregnancy-related AKI (PR-AKI) in terms of KDIGO and AKIN criteria [3]. The inclusion of new definitions has led to increased recognition of PR-AKI and thereby, increased incidence. This fact is evident from the observation of an increased incidence of PR-AKI from a recent report from Canada. The incidence of PR-AKI has increased in Canada from a mere 1.66 per 10,000 deliveries between 2003 and 2004 to 2.68 per 10,000 deliveries between 2009 and 2010.

Etiology

Majority of the cases of PR-AKI occur during the third trimester. Preeclampsia is the most common cause of PR-AKI. Other causes include—septic abortion, hemorrhagic complications—antepartum hemorrhage (APH) or postpartum hemorrhage (PPH), and thrombotic microangiopathies (TMAs).

Causes of AKI during pregnancy	Trimester	AKI
Obstetric complications		
Septic abortion (20%–30%)	First	Intrarenal
Antepartum hemorrhage—placenta praevia, abruptio	Third	Prerenal
Placentae		
Postpartum hemorrhage—puerperal sepsis	Third	Prerenal
Intrauterine fetal death (25%)	Third	Intrarenal
Pregnancy specific causes		
Preeclampsia/eclampsia (30%–40% most common cause)	Second/third	Intrarenal
HELLP syndrome AFLP (2%–4%)	Third	Intrarenal
Atypical hemolytic uremic syndrome and thrombotic thrombocytopenic purpura	Postpartum	Prerenal and intrarenal
Hyperemesis gravidarum and other causes	First, second, or third	Postrenal
Lupus nephritis	First	
Hydronephrosis	First, second, or third	
	Third	

Preeclampsia is the most common cause of renal disease in pregnancy [4]. About 30%–40% of patients with PR-AKI have preeclampsia. Presence of decreased platelet counts in preeclamptic patients should raise a clinical suspicion of HELLP syndrome.

TMAs are a group of rare disorders characterized by the presence of platelet fibrin thrombi in vasculature leading to platelet consumption and microangiopathic hemolytic anemia (MAHA).

Pyelonephritis and asymptomatic bacteriuria have high incidence in pregnant patients. Presence of asymptomatic bacteriuria increases the incidence of preterm births and low birth weight babies.

Management of renal dysfunction in pregnancy

Clinical management of pregnant patients with renal failure requires an entire team of multidisciplinary doctors—obstetricians and nephrologists. Initial patient stabilization and optimal management are key to successful management. Safety of both the mother and fetus should always be borne in mind.

1. **Stabilize the patient**

   For bleeding patients with placenta previa or abruption, ongoing bleeding areas should be secured and the underlying cause should be treated.

2. **Protect kidneys from further damage**

   Patients should undergo hemodynamic monitoring to assess the patient for fluid responsiveness. Crystalloids are preferred over colloids. Colloids increase the chances of AKI and should be avoided. Among crystalloids, balanced salt solutions are preferred over normal saline.

3. **Symptomatic management**

   Hyperkalemia—Prompt management with glucose insulin drip, salbutamol nebulization, potassium-binding resin, and diuretics. For refractory cases, urgent dialysis is required.

   Metabolic acidosis—Cases of severe metabolic acidosis ($pH < 7.15$, $HCO3 < 15$ mmol/L) should be managed with intravenous bicarbonate infusion.

4. **Renal replacement therapy**

   Renal replacement therapy is the only definitive treatment for renal dysfunction in pregnancy. RRT includes—peritoneal dialysis (PD), nocturnal hemodialysis, intermittent hemodialysis (IHD), and continuous renal replacement therapy (CRRT).

   There are no major differences between the efficacy of the two modalities (IHD and CRRT) in terms of mortality and renal function recovery data. Whereas IHD operates principally by the process of diffusion, the CRRT utilizes ultrafiltration and convection to remove lot of extra solvents along with the solutes (solvent drag). It also removes the middle molecules primarily responsible for sepsis. IHD and CRRT also differ in the blood flow rate. In CRRT, blood flow rate may be as low as 150 mL/min, while in IHD, blood flow rates are in the range of 300−450 mL/min.

   CRRT is clearly superior in terms of better hemodynamic stability, enhanced solute clearance, and fluid balance. CRRT avoids intradialytic hypotension, which might reduce the placental perfusion pressure and may be deleterious to the developing fetus.

Continuous renal replacement therapy

CRRT is the modality of choice for the management of renal dysfunction in pregnancy. CRRT should be offered in cases of severe volume overload, refractory metabolic acidosis, severe and refractory hyperkalemia, uremic encephalopathy, uremic pericarditis, etc. CRRT is the preferred modality for hemodynamically unstable patients. CRRT is advantageous in terms of superior solute clearance, hemodynamic stability, and better management of water and electrolytes. Currently, it is widely used in pregnant patients and is the mainstay of management of renal dysfunction in pregnancy. However, the use of CCRT in pregnancy is different as a major change in dialysis regimen is often deemed necessary.

Basic principles diffusion

Diffusion is defined as the movement of a solute from an area of high concentration to an area of low concentration across a semipermeable membrane. As weight of solute increases, diffusion across the semipermeable membrane reduces. This is the primary mechanism involved in continuous venovenous hemodialysis (CVVHD).

Fig. 16.1 depicts a schematic representation of diffusion.

Convection

When solute movement is against pressure gradient, it is called convection. In this process, a lot of solute is often dragged with the solvent leading to efficient solvent removal (solvent drag). Convection-based modalities like continuous venovenous hemofiltration (CVVHF) and continuous venovenous hemodiafiltration (CVVHDF) use high flux filters and are efficient in removing middle and large molecular weight solutes.

Fig. 16.2 demonstrates the basic principle underlying convection. Human kidneys work on the principle of convection making the process more physiological.

Different modalities of CCRT

There are several different types of CRRT used clinically in pregnant patients with renal dysfunction. They are classified on the basis of dialyzate and ultrafiltrate flow patterns. They are as follows.

1. Slow continuous ultrafiltration (SCUF)
2. CVVHD
3. CVVHF
4. CVVHDF

Slow continuous ultrafiltration

As is clear from its name, SCUF aims at safe fluid removal as the primary goal. SCUF is particularly helpful in cases of preeclampsia with diuretic resistant and hemodynamically unstable pulmonary edema. Fig. 16.3 depicts removal of ultrafiltrate in SCUF.

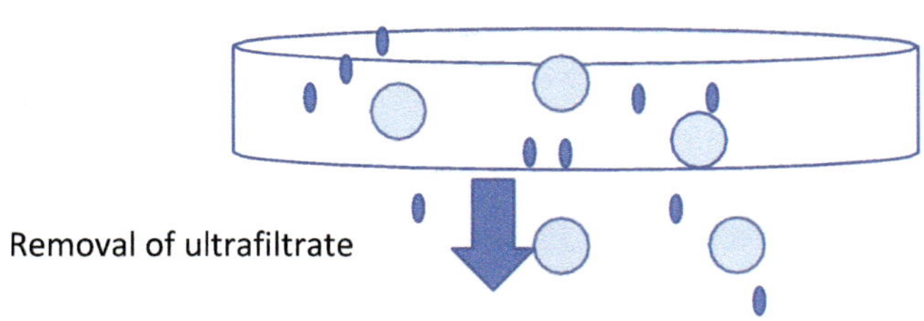

Figure 16.2 Convection—removal of both small and large molecular weight molecules.

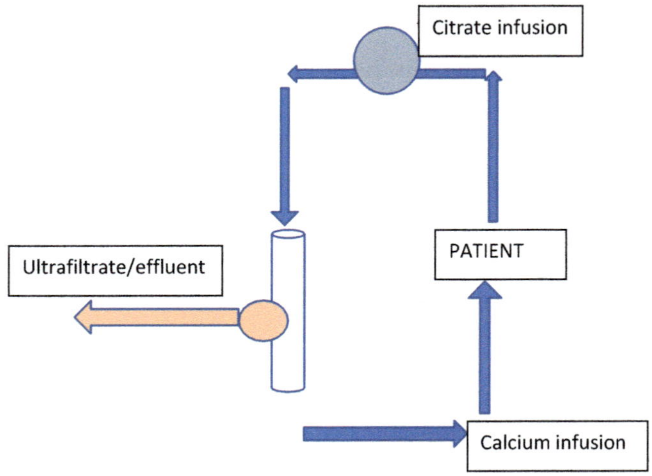

Figure 16.3 Slow continuous ultrafiltration (SCUF).

Continuous venovenous hemofiltration

Since it incorporates convection as the primary modality, it aims to remove large molecular weight solutes along with small molecular weight solutes. Hence, it ensures safe management of fluid volume along with solute removal. Also, it requires administration of replacement solution to compensate for the loss of large volumes of ultrafiltrate. This is shown in Fig. 16.4. CVVH is primarily indicated in pregnant patients with AKI, severe metabolic and electrolyte abnormalities.

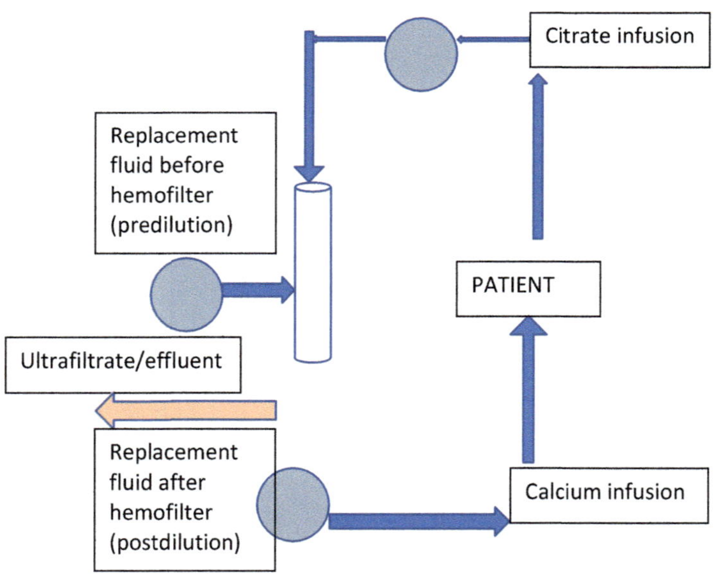

Figure 16.4 Continuous venovenous hemofiltration (CVVH).

Continuous venovenous hemodialysis

The primary principle behind CVVHD is diffusion and it aims for fluid and solute removal (primarily small molecular weight molecules). Dialyzate is perfused in direction opposite to blood flow across hemofilter and diffusion of solutes takes place. Fig. 16.5 demonstrates CVVHD.

Continuous venovenous hemodiafiltration

CVVHDF combines the principles of both diffusion and convection. It requires both dialyzate and replacement fluid administration. CVVHDF is helpful in the removal of small, medium, and large molecular weight solutes. Fig. 16.6 shows replacement fluid and dialyzate administration in CVVHDF.

Components of RRT

The components of CRRT are vascular access, hemofilter, anticoagulation, and CRRT system. They are discussed in detail as follows.

Vascular access

Central venous catheters are often secured for starting RRT as a basic prerequisite. Commonly chosen sites are internal jugular vein, subclavian, and femoral vein. For securing femoral access, an extra-long catheter (>20 cm) may be necessary.

Tunneled catheters (permacath) may also be placed for patients requiring long-term dialysis.

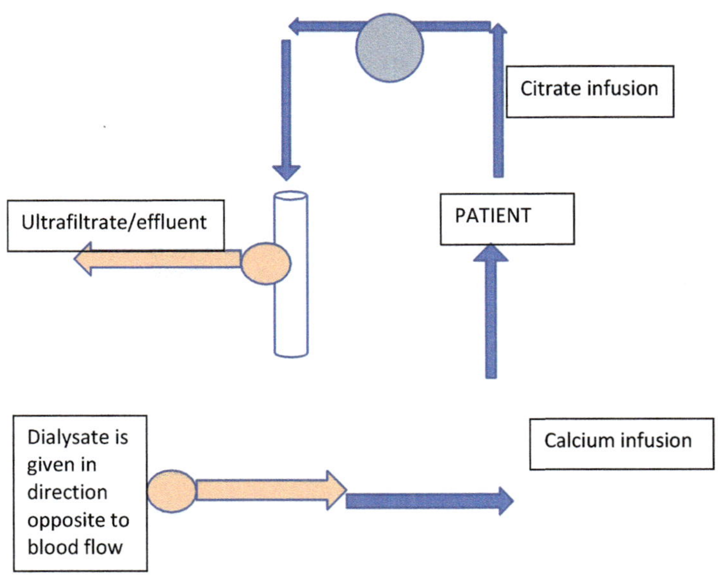

Figure 16.5 Continuous venovenous hemodialysis (CVVHD).

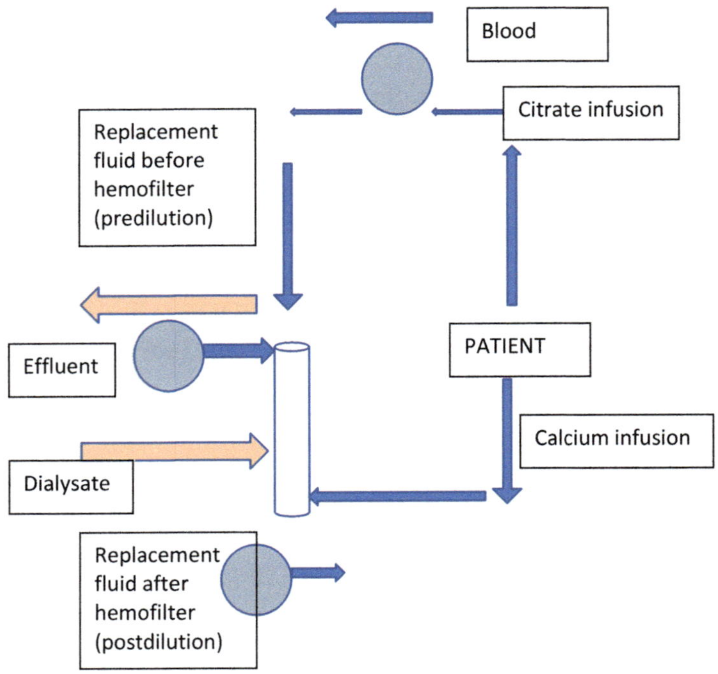

Figure 16.6 Continuous veno-venous hemodiafiltration (CVVHDF).

Anticoagulation

Anticoagulation prevents clotting of blood in the circuit and keeps it patent during the dialysis. Low dose heparin is usually administered prefilter in a dose of 5–10 U/kg/hr. Heparin is considered safe in pregnancy as it does not cross placental barrier. Routine monitoring with aPTT is required as an essential part of the procedure. However, its use should be limited in patients with active bleeding and thrombocytopenia.

CRRT system (CRRT machine, line sets, membrane, hemofilter, etc.)

Semipermeable membrane required for diffusion and convection is present in the hemofilter. Uncoated AN-69 membranes may lead to release of bradykinin and hypotension known as bradykinin release syndrome. Polyethyleneamine coating on AN-69 membranes avoid this syndrome.

Trained intensive care team

An absolute necessity is staff trained in dealing with CRRT machine.

Replacement fluid

Since dialysis removes large volume of fluid by convection, it should be replaced appropriately. Replacement can be done either prefilter or postfilter. Bicarbonate, acetate, or lactate are used as a buffer.

Indications of starting CRRT

Patients presenting with renal dysfunction and having hyperkalemia, pulmonary edema, volume overload, blood urea nitrogen (BUN >100 mg/dL), uremic pericarditis, and uremic encephalopathy warrant renal replacement therapy. The indications of starting CRRT are depicted in Table 16.2.

Timing of initiation of CRRT

The optimal timing of initiation of CRRT has often been a matter of debate. Majority of studies and clinical trials have been done in nonpregnant patients with AKI. Early provision of CRRT is beneficial as it removes inflammatory mediators—IL-1, IL-6, and TNF-alfa, provides prompt correction of metabolic complications, and also prevents volume overload. In ELAIN trial, 231 postoperative cardiac surgical patients were compared in terms of provision of early RRT (KDIGO stage 2), without uremic complications as compared to late therapy (KDIGO stage 3) [5]. Early CRRT was found to lower the 90-day mortality. However, clinical trials (AKIKI, STARRT-AKI, and IDEAL—ICU trial) done subsequently failed to conclude the same [6,7]. Mortality outcomes were similar between early and delayed RRT groups in most of these trials. They also found RRT was largely unnecessary in early group and it led to the development of complications like hypotension and catheter-related blood stream infection (CRBSI) [6,7]. So, consensus is in favor of provision of late RRT in nonpregnant patients with renal dysfunction.

However, since pregnancy is a state of augmented renal clearance, the usual timing and intensity of hemodialysis needs to be intensified and accelerated for optimal urea clearance goals. It is hypothesized that intrauterine azotemic environment can have detrimental effects on the developing fetus and may worsen fetal outcome. In a small study done by Banerjee et al., fetal outcomes were found to be worse in pregnant women undergoing late dialysis [8]. Typically, ultrafiltration goal of one to two liters per session of dialysis is considered adequate.

Intensity of CRRT in pregnancy

Increasing frequency of dialysis and/or providing dialysis for prolonged periods reduces polyhydramnios, controls hypertension, improves baby birth weight, and increases

Table 16.2 Indications of starting CRRT.

S. No.	Renal indications
1.	Pulmonary edema
2.	Acute metabolic acidosis in sepsis
3.	Diuretic resistant hyperkalemia
4.	Refractory/diuretic resistant volume overload
5.	AKI in septic shock, KDIGO stage 2/3

chances of live birth [9,10]. Increased dialysis time (>20 h/week) is associated with better fetal outcomes. The optimal dose of dialysis depends on residual renal function.

Most nephrologists would like to increase the frequency of dialysis to five to six times per week. In a study performed in Canada and United States, extended session dialysis for 36 h/week was compared with standard sessions of 17 h/week. The authors observed higher mean birth weights among pregnant patients receiving 36 h dialysis sessions [11].

Although, there are case reports of successful pregnancies in patients with renal dysfunction undergoing usual/standard dialysis regimen, it is prudent to consider increasing the dose of dialysis in pregnant patients.

Inadequate dialysis often leads to polyhydramnios and premature births.

There are very few studies published till date comparing standard versus high intensity dialysis dose. However, several factors make it prudent to consider increasing the dose and frequency of dialysis in pregnancy and consider predialysis urea of around 100 mg/dL. Studies have shown that preterm delivery and intrauterine growth retardation correlate with the uremia [12]. In a study done by Bagon et al., a clear correlation was found between increasing the dose of dialysis and birth weights. Bagon [13] et al. increased the dose of dialysis to a weekly Kt/V of 6 to 8. They observed that increased dose of dialysis is associated with better fetal outcomes. Increasing dialysis dose reduces polyhydramnios and lowers urea load.

Also, frequent dialysis reduces the incidence of intradialysis hypotension which is potentially deleterious to the developing fetus.

Special concerns

High dose and intensity of dialysis in pregnancy requires careful monitoring of potassium and phosphorus, which may be removed in excessive amounts during dialysis sessions.

Avoid intradialysis hypotension

Close monitoring should be done for pregnant patients undergoing dialysis. Uteroplacental circulation might be compromised due to hypotension. This is especially important during the last portion of the dialysis session.

Assessment of volume status weekly

A careful general physical examination of pregnant patients should be performed to assess for volume status. Presence of hypertension and increase in dry weight point toward fluid overload. Difficulty may arise in differentiating between volume overload and pregnancy-associated weight gain.

Potassium

Most nephrologists prefer using a dialyzate with potassium >3 mEq/L. This may be helpful in reducing risks associated with hypokalemia. Hyperkalemia is an uncommon sequelae in pregnant patients with renal dysfunction owing to high intensity dialysis.

Phosphate

Hypophosphatemia may develop despite a good phosphate intake in pregnant patients undergoing high intensity dialysis. Dialyzate may be enriched with sodium phosphate to aim for a dialyzate phosphate concentration of up to 4.5 mg/dL.

Hyperphosphatemia is rare in pregnant patients.

Bicarbonate

Dialyzate concentration of 30 mEq/L is advised to avoid metabolic acidosis. This is especially necessary in patients undergoing intensive dialysis regimen. In case of metabolic acidosis in patient, dialyzate concentration may be increased to 35 mEq/L.

Water-soluble vitamins

Pregnant patients undergoing dialysis in pregnancy require increased doses of water-soluble vitamins sp. folate. The requirements may go up to double of the usual dose. Since folate is cleared by dialysis, it is recommended that folic acid supplementation of 5 mg/day is essential in these patients.

Nutrition

Special attention should be given to pregnant patients on high intensity dialysis as it makes the patients more prone to depletion syndrome. Dietary prescription should be modified accordingly.

Daily protein intake of 1.5–1.8 g/kg/day

Pregnant patients on dialysis often require a higher than usual amount of protein supplementation in their diets to ensure replenishment of amino acid loss during dialysis.

Erythropoiesis-stimulating agents

Higher doses of erythropoiesis-stimulating agents (ESAs) may be necessary to maintain an adequate red cell mass. Most of the ESAs do not cross the placental barrier and are deemed safe in pregnancy.

Care should be taken in maintain adequate iron stores also. Parenteral iron may be supplemented when necessary.

Anticoagulation

The anticoagulant of choice in pregnancy is heparin.

Renal transplantation

Renal transplantation provides a fair chance of parenthood and survival to pregnant patients with renal dysfunction. Advantages of renal transplantation is successful pregnancies and lesser birth anomalies.

Fig. 16.7 depicts highlights of CRTT in pregnancy.

Summary

1. Physiological changes of pregnancy in kidneys include an augmented renal clearance (raised GFR and RPF) and low levels of urea and creatinine.
2. Pregnant patients undergoing dialysis should be monitored by both a nephrologist and a maternal–fetal medicine specialist.
3. CRRT is the modality of choice as it offers superior solute clearance and offers hemodynamic stability.
4. Dose (>20 h/week) and intensity of dialysis (5–6 times a week) should be increased to account for augmented renal clearance of solutes in pregnancy.

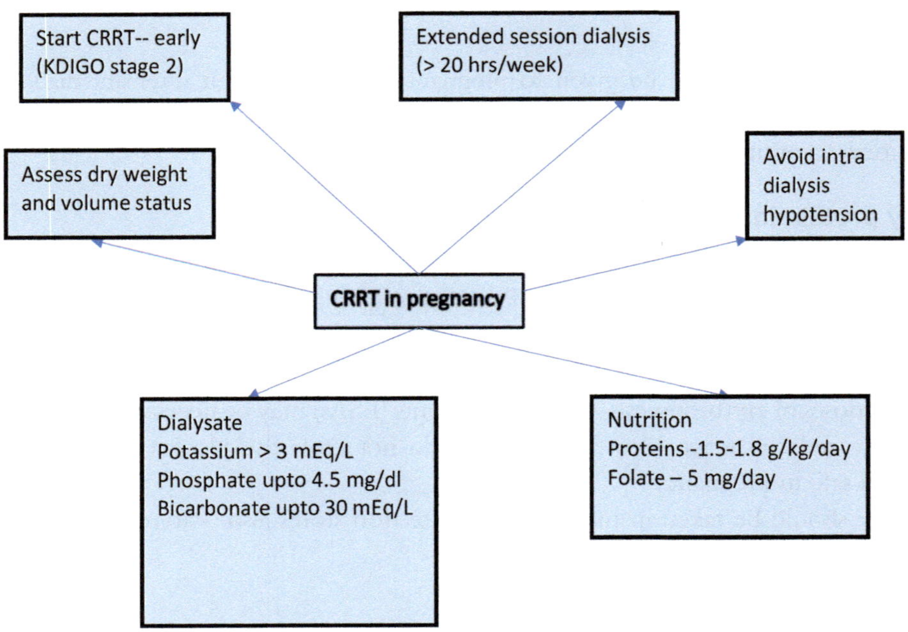

Figure 16.7 Schematic representation of CRRT in pregnancy.

5. Optimum monitoring of dry weight and assessment of volume overload should be done to ensure optimum dialysis.
6. Special care should be taken to ensure good nutrition and avoid electrolyte imbalances in case of high intensity dialysis sessions.

References

[1] Maynard SE, Thadhani R. Pregnancy and the kidney. J Am Soc Nephrol January 2009;20(1):14–22. https://doi.org/10.1681/ASN.2008050493. Epub 2008 Dec 31. PMID: 19118150.

[2] Stratta P, Besso L, Canavese C, Grill A, Todros T, Benedetto C, et al. Is pregnancy- related acute renal failure a disappearing clinical entity? Ren Fail 1996;18:575–84.

[3] Mehrabadi A, Liu S, Bartholomew S, et al. Hypertensive disorders of pregnancy and the recent increase in obstetric acute renal failure in Canada: population based retrospective cohort study. BMJ 2014;349:g4731.

[4] Prakash J, Ganiger VC. Acute kidney injury in pregnancy-specific disorders. Indian J Nephrol 2017; 27(4):258–70. https://doi.org/10.4103/0971-4065.202406. PMID: 28761227; PMCID: PMC5514821.

[5] Zarbock A, Kellum JA, Schmidt C, Van Aken H, Wempe C, Pavenstädt H, Boanta A, Gerß J, Meersch M. Effect of early vs delayed initiation of renal replacement therapy on mortality in critically ill patients with acute kidney injury: the ELAIN randomized clinical trial. JAMA May 24–31, 2016; 315(20):2190–9. https://doi.org/10.1001/jama.2016.5828. PMID: 27209269.

[6] Barbar SD, Clere-Jehl R, Bourredjem A, Hernu R, Montini F, Bruyère R, Lebert C, Bohé J, Badie J, Eraldi JP, Rigaud JP, Levy B, Siami S, Louis G, Bouadma L, Constantin JM, Mercier E, Klouche K, du Cheyron D, Piton G, Annane D, Jaber S, van der Linden T, Blasco G, Mira JP, Schwebel C, Chimot L, Guiot P, Nay MA, Meziani F, Helms J, Roger C, Louart B, Trusson R, Dargent A, Binquet C, Quenot JP, IDEAL-ICU Trial Investigators and the CRICS TRIGGERSEP Network. Timing of renal-replacement therapy in patients with acute kidney injury and sepsis. N Engl J Med October 11, 2018;379(15):1431–42. https://doi.org/10.1056/NEJMoa1803213. PMID: 30304656.

[7] Zampieri FG, da Costa BR, Vaara ST, Lamontagne F, Rochwerg B, Nichol AD, McGuinness S, McAuley DF, Ostermann M, Wald R, Bagshaw SM, STARRT-AKI Investigators. A Bayesian reanalysis of the standard versus accelerated initiation of renal-replacement therapy in acute kidney injury (STARRT-AKI) trial. Crit Care August 25, 2022;26(1):255. https://doi.org/10.1186/s13054-022-04120-y. PMID: 36008827; PMCID: PMC9404618.

[8] Banerjee A, Mehrotra G. Comparison of standard conservative treatment and early initiation of renal replacement therapy in pregnancy-related acute kidney injury: a single-center prospective study. Indian J Crit Care Med August 2020;24(8):688–94. https://doi.org/10.5005/jp-journals-10071-23537. PMID: 33024376; PMCID: PMC7519595.

[9] Piccoli GB, Conijn A, Consiglio V, Vasario E, Attini R, Deagostini MC, Bontempo S, Todros T. Pregnancy in dialysis patients: is the evidence strong enough to lead us to change our counseling policy? Clin J Am Soc Nephrol January 2010;5(1):62–71. https://doi.org/10.2215/CJN.05660809. Epub 2009 Nov 5. PMID: 19965547; PMCID: PMC2801656.

[10] Asamiya Y, Otsubo S, Matsuda Y, Kimata N, Kikuchi KAN, Miwa N, Uchida K, Mineshima M, Mitani M, Ohta H, Nitta K, Akiba T. The importance of low blood urea nitrogen levels in pregnant patients undergoing hemodialysis to optimize birth weight and gestational age. Kidney Int June 2009; 75(11):1217–22. https://doi.org/10.1038/ki.2009.48. Epub 2009 Feb 25. PMID: 19242506.

[11] Hladunewich MA, Hou S, Odutayo A, Cornelis T, Pierratos A, Goldstein M, Tennankore K, Keunen J, Hui D, Chan CT. Intensive hemodialysis associates with improved pregnancy outcomes: a Canadian and United States cohort comparison. J Am Soc Nephrol May 2014;25(5):1103–9. https://doi.org/10.1681/ASN.2013080825. Epub 2014 Feb 13. PMID: 24525032; PMCID: PMC4005313.

[12] Jones DC, Hayslett JP. Outcome of pregnancy in women with moderate or severe renal insufficiency. N Engl J Med 1996;335:226–32.
[13] Bagon JA, Vernaeve H, De Muylder X, Lafontaine JJ, Martens J, Van Roost G. Pregnancy and dialysis. Am J Kidney Dis May 1998;31(5):756–65. https://doi.org/10.1016/s0272-6386(98)70060-5. PMID: 9590184.

Further reading

[1] Hull AR. More dialysis appears beneficial for pregnant ESRD patients (at least in Belgium). Am J Kidney Dis May 1998;31(5):863–4. https://doi.org/10.1016/s0272- 6386(98)70059-9. PMID: 9590200.

CHAPTER 17

Pregnancy and chronic kidney disease

Hythem Al-Sum, Maisah Albakri and Suhad Alsurori
King Abdulaziz Medical City, Riyadh, Saudi Arabia

Introduction

The prevalence of chronic kidney disease (CKD) in the general population is increased in the past years and affects almost 4% of women of childbearing age [1]. With this increase in the incidence of CKD, the shift to delay pregnancy, and improvements in the care of these patients, the incidence of CKD in pregnancy has become more common.

Pregnancy in women with CKD of any stage poses clinical challenges associated with many adverse maternal and perinatal outcomes including future impact on maternal renal health, preeclampsia, fetal growth restriction, and preterm delivery. Women with early stage CKD have a high likelihood of successful pregnancy and are less likely to develop deterioration in renal function during pregnancy. On the other hand, pregnant women with advanced CKD have an increased risk of accelerated and irreversible decline in renal function and poor outcomes in pregnancy [2]. Therefore, the management of women with CKD in pregnancy needs to be shared between specialized obstetric and nephrology care to minimize the adverse effect of pregnancy on maternal renal function and the subsequent effects on the fetus.

Pregnancy in women with CKD

Definition of chronic kidney disease

CKD is defined as kidney damage or glomerular filtration rate (GFR) <60 mL/min/1.73 m^2 [3] for 3 months or more, irrespective of diagnosis. Kidney damage can be established by the presence of albuminuria, albumin-to-creatinine ratio >30 mg/g in two of three urine samples (Table 17.1).

Stages of chronic kidney disease

Severity of kidney disease is classified into five stages according to the level of GFR, with higher stages representing lower GFR levels (Table 17.2).

Preconception counseling

Although CKD can present during pregnancy, its presence is usually known before conception. Those women with known CKD should have preconception counseling to discuss the benefits of careful pregnancy planning and discuss contraceptive options

The Kidney of the Critically Ill Pregnant Woman
ISBN 978-0-443-21473-8
https://doi.org/10.1016/B978-0-443-21473-8.00007-0

Table 17.1 Definition of chronic kidney disease criteria [4].

1. Kidney damage for >3 months, as defined by structural or functional abnormalities of the kidney, with or without decreased GFR, manifest by either:
   - Pathological abnormalities; or
   - Markers of kidney damage, including abnormalities in the composition of the blood or urine, or abnormalities in imaging tests
2. GFR <60 mL/min/1.73m^2 for >3 months, with or without kidney damage.

Table 17.2 Stages of chronic kidney disease [4].

Stage	Description	GFR (mL/min/1.73m^2)
1	Kidney damage with normal or increased GFR	≥90
2	Kidney damage with mild decrease in GFR	60–89
3	Moderate decrease in GFR	30–59
4	Severe decrease in GFR	15–29
5	Kidney failure	<15 (or dialysis)

if pregnancy needs to be delayed. Specialized multidisciplinary team care including obstetricians, nephrologists, anesthetists, and dieticians is often beneficial for women with CKD [2,3].

Key areas for discussion include an exploration of risks and outcomes as described below, assessment of associated risk factors (including control of primary renal disease, blood pressure, and proteinuria); adjustment of medication doses or change to medications that are safe during pregnancy (teratogenic medications—such as angiotensin-converting enzyme inhibitors (ACEIs) and angiotensin II receptor blockers—should be stopped before pregnancy if equally effective drugs are available, or as soon as pregnancy is confirmed); decisions making regarding the time of conception; and development of a pregnancy, delivery, and postpartum management plan.

Genetic counseling should be offered for those who have an inheritable renal disease such as polycystic kidney disease. As for all pregnant women, in preconception and first trimester (first 12 weeks of gestation) prophylactic folic acid 400 mcg/day supplementation is recommended, but if the woman is on dialysis, 3–5 mg/day may be required, as the folic acid will be cleared during dialysis. Low-dose aspirin (75–150 mg/day) should be started in early pregnancy to reduce the risk of preeclampsia and improve perinatal outcomes if they are thought to be important for protecting maternal renal function. While the overall risk of pregnancy problems is increased in women with CKD, many women will have uncomplicated pregnancies and excellent outcomes, but still require frequent antenatal review, particularly monitoring maternal renal function and blood pressure, and fetal growth and well-being.

Women with known CKD and renal function impairment should be informed that the renal disease might progress during pregnancy, especially in those with moderate-to-severe disease [5]. Most pregnancies associated with moderate-to-severe renal insufficiency will result in a premature birth [6]. Moreover, women with a serum creatinine level greater than 2.0 mg/dL or GFR less than 30 mL/min/1.73 m^2 should be counseled that they have a one-in-three chance of progressing to end-stage renal disease (ESRD) within 1 year postpartum.

In the presence of nephrotic syndrome, pregnancy should be delayed until appropriate treatment has been given [5,7]. ACEIs and angiotensin receptor blockers (ARBs) should be avoided, if possible, before conception.

Women with diabetes should undergo a thorough evaluation of their nephropathy and achieve optimal glycemic control before becoming pregnant. Pregnancy should be discouraged in women with diabetes nephropathy who have a serum creatinine above 1.5—1.7 mg/dL or a GFR <60 mL/min/1.73 m^2, especially in the presence of uncontrolled hypertension [8].

In women with glomerulonephritis including lupus nephritis, conception should preferably be planned for a period of stable, sustained remission of at least 6 months [9]. Immunosuppression should be adjusted and continued throughout pregnancy to decrease the risk of disease flare. Corticosteroids and azathioprine can be used safely during pregnancy since they are not associated with major developmental abnormalities [10]. Cyclosporine does not appear to be a major teratogen, as suggested by a metaanalysis of 15 studies [11]. Registry data suggest that dose adjustments are required in the majority of pregnant women who receive cyclosporine to maintain therapeutic levels established before conception as pregnancy is associated with alterations in cyclosporine metabolism and distribution [12]. There is a paucity of data concerning the effect of tacrolimus on pregnancy. As with cyclosporine, patients taking tacrolimus require frequent monitoring of drug levels [8]. Mycophenolate mofetil (MMF) is contraindicated in pregnancy. According to the European Best Practice Guidelines, MMF should be discontinued at least 6 weeks before attempted conception and patients switched to another agent (such as azathioprine) [13].

Effect of pregnancy on CKD

The risk of renal deterioration during pregnancy remains uncertain as most studies have different outcomes [14]. A metaanalysis by Zhang et al. [14] showed no association between pregnancy and deterioration of renal function, regardless of etiology or prepregnancy stage, though few included studies had advanced CKD patients. Piccoli et al. [15] observed 7.6% of women with CKD stage 1 experienced a shift in the stage or start on renal replacement therapy with pregnancy, worsening by 20% to stage 4—5. Jones and

Hayslett [9] in their study of renal function in women with moderate-to-severe CKD with a creatinine level of over 124 μmol/L during pregnancy and postpartum found transient deterioration in 43% and permanent deterioration in 31% of cases. Similarly, a small but widely cited study by Imbasciati et al. [16] found prepregnancy eGFR less than 40 mL/min/1.73 m^2 with proteinuria more than 1 g/24 h predicted renal decline after pregnancy. Therefore, evidence that pregnancy worsens the early stage of CKD is limited and uncertain, though, in advanced stages, the evidence is strong.

Chronic kidney disease: Stages 1 and 2

If blood pressure controlled and proteinuria is minimal (<1 g per 24 h), in general, they have a good outcome and successful pregnancies with no long-term consequences [17]. One study analyzed the outcomes of 360 women with chronic glomerulonephritis and normal renal function over a 30-year period: 171 became pregnant while 189 did not. After an average of 15 years (and up to 30 years) of follow-up, no difference in the development of ESRD was noted between the two group [18]. Although the long-term renal prognosis for women with stages 1 and 2 is reassuring, these pregnancies are at increased risk for developing preeclampsia and intrauterine fetal growth restriction. One study of 46 women with stage 1 in pregnancy reported a 39% occurrence of preeclampsia and 17% of infants who are small for gestational age (SGA) [7]. A recent study compared pregnancy outcomes among 8900 women with no history of AKI and a serum creatinine available within 6 months of starting pregnancy, to 105 women who had recovered from AKI (R-AK) [19]. Both groups had normal serum creatinine values (~0.7 mg/dL) preconceptions: for the r-AKI group, 23% risk for preeclampsia as compared with a 4% risk (aOR 6.5, $P < 0.001$) in the no-AKI group. Similarly, the risk for small neonates was 15% in the r-AKI group compared with 9% in the controls (aOR 2, $P = 0.02$). This result showed a new risk factor for preeclampsia and associated complications; however, the mechanisms remain poorly understood.

Chronic kidney disease: Stages 3 through 5

The renal outcome for pregnant patients with stages 3–5 CKD is less favorable. One of the largest studies published in 1996 about renal outcomes in pregnant women with CKD included 82 pregnancies in 67 women with primary renal disease and a serum creatinine level of 1.4 mg/dL or greater, either before conception or in the first trimester, GFR was stable 51% during pregnancy and 6 months after delivery, decreased but returned to prepregnancy levels in 8%, decreased and persisted lower at 6 months in 31%, and decreased between 6 weeks and 6 months after delivery in 10%. The likelihood of an accelerated decline in renal function was greater in women whose first measured serum creatinine was greater than 2 mg/dL. Only 1 out of 49 women with a serum creatinine level of 1.4–1.9 mg/dL experienced an accelerated decline, as opposed to 7 out of 21 pregnancies in women with a baseline serum creatinine level of 2 mg/dL or greater.

Eight women from the entire group progressed to ESRD within 1 year of pregnancy [9]. A more recent study of 49 women with a mean serum creatinine level of 2.1 mg/dL and mean GFR of 35 mL/min/1.73 m^2 at the time of conception demonstrated that both presences of a baseline GFR of less than 40 mL/min/1.73 m^2 and proteinuria more than 1 g/d predicted increased in GFR loss after delivery compared with before conception. The presence of both risk factors also reduces GFR by 50% more quickly and can even prevent initiating dialysis [20].

A recent study published in 2015 for 64 pregnant women found that pregnancy and delivery did not show any significant changes in the renal function of patients with stage 3 CKD with $\geq$45 mL/min at 5 years after delivery, although proteinuria was increased around 30 weeks of pregnancy up to 3 months after delivery compared to patients with stage 3 CKD with <45 mL/min, there was a significant deterioration in the GFR at 5 years after delivery. Additionally, pregnant women with stage 3 CKD were compared with 22 nonpregnant women with similar clinical and demographic characteristics [21].

Effect of CKD on pregnancy

The CKD stage has a greater effect on the outcome than the etiology of the kidney disease. An exception is systemic lupus erythematosus (SLE) nephritis: outcomes are poorer than for women with the same renal impairment from other causes, particularly if associated with antiphospholipid antibodies. Regarding lupus nephritis, a recent metaanalysis identified risks of SLE flare (25.6%), hypertension (16.3%), preeclampsia (7.6%), premature birth (39.4%), and intrauterine growth restriction (12.7%) in 37 studies of 2751 pregnancies [22]. Differentiating preeclampsia from SLE nephritis is difficult. Key differentiating features include the typical rash, arthritis, hematuria with cellular casts, low C3/C4 complement levels, or the presence of antiphospholipid antibodies in SLE, while abnormal liver function tests suggest preeclampsia rather than SLE flare [5,23].

Diabetic patients have increased rates of preeclampsia, fetal abnormalities, and fetal loss; proteinuric diabetic nephropathy may worsen in pregnancy, and the combination of diabetes, proteinuria, renal impairment, and hypertension is particularly difficult [8]. These risks may be elevated in women with "overt" diabetic nephropathy compared with those with early stage disease [9]. Establishing good glycemic control and lowering the hemoglobin A1c preconception is important to improve maternal and neonatal outcomes, including reducing the risk of fetal malformation. Some evidence indicates the risk of congenital abnormalities in diabetic women may be increased in the presence of nephropathy, underscoring the need for optimized glycemic control [23].

Women with autosomal dominant polycystic kidney disease have an increased risk of preeclampsia and urinary tract infection [5,24]. Hepatic cysts may increase in the estrogenic environment and known cerebral aneurysms should be assessed before labor [6,23]. There is also a risk of intracity hemorrhage during labor, and large renal cysts should be monitored before and following delivery [25]. Outcomes in other common nephropathies including IgA nephropathy do not consistently differ from same-stage CKD of other etiology [26], though robust comparative data between primary diseases remain limited.

In the aspect of obstetric outcomes, pregnant women with moderate-to-severe kidney disease have increased risks for preeclampsia (>70%), moderate-to-severe anemia (>60%), and SGA neonates (>50%), and prematurity in 6% of pregnancies delivered before 28 weeks, 24% of pregnancies delivered before 34 weeks, and 76% of pregnancies delivered before full term (37 weeks' gestation) [27]. There are no documented fetal congenital anomalies from CKD itself, the congenital anomalies are caused because of the teratogenic effect of some medications, as noted in the table [28].

Management of pregnancy and complications in women with CKD

The following algorithm summarizes the recommended approach to care for women with CKD in pregnancy. Additional management points are also discussed (Fig. 17.1).

Evaluation of renal function and proteinuria

Proteinuria normally increases in pregnancy due to its physiological change by increased GFR. The upper limit of normal proteinuria is up to 300 mg/24 h or 30 mg/μmol creatinine [24], 30% of nonproteinuric women will develop proteinuria during pregnancy [29]. If urine protein reaches the nephrotic range (more than 3 g/24 h), with low serum albumin of less than 30 g/dL, it is advised that women should be started on thromboprophylaxis [30], The doses should be adjusted with a clinical pharmacist according to the renal impairment level.

Management of primary disease

Pregnant women with CKD should have regular follow-ups with nephrologists along with fetal-maternal medicine specialists to ensure the stability of renal function and promote the quiescence of disease and optimize short and long-term health outcomes [31].

Observational data published in 2015 and 2017 about the association between active lupus nephritis, nephrotic syndrome, and small vessel vasculitis with increased risks of adverse pregnancy outcomes including fetal death [26,32] showed that stable renal function is associated with better pregnancy outcomes [33]. Disease stability should be ensured before conception [34].

Specific diseases, such as lupus nephritis, need to be stable with no flare and immunosuppressive medication that is safe during pregnancy for at least 6 months before conception. Regarding tacrolimus, a recent study showed that it can be used during pregnancy, in addition, anti-Ro and anti-La need to be done as the fetus will be at risk for congenital heart block and neonate lupus [35]. Those who test positive for these antibodies should continue hydroxychloroquine.

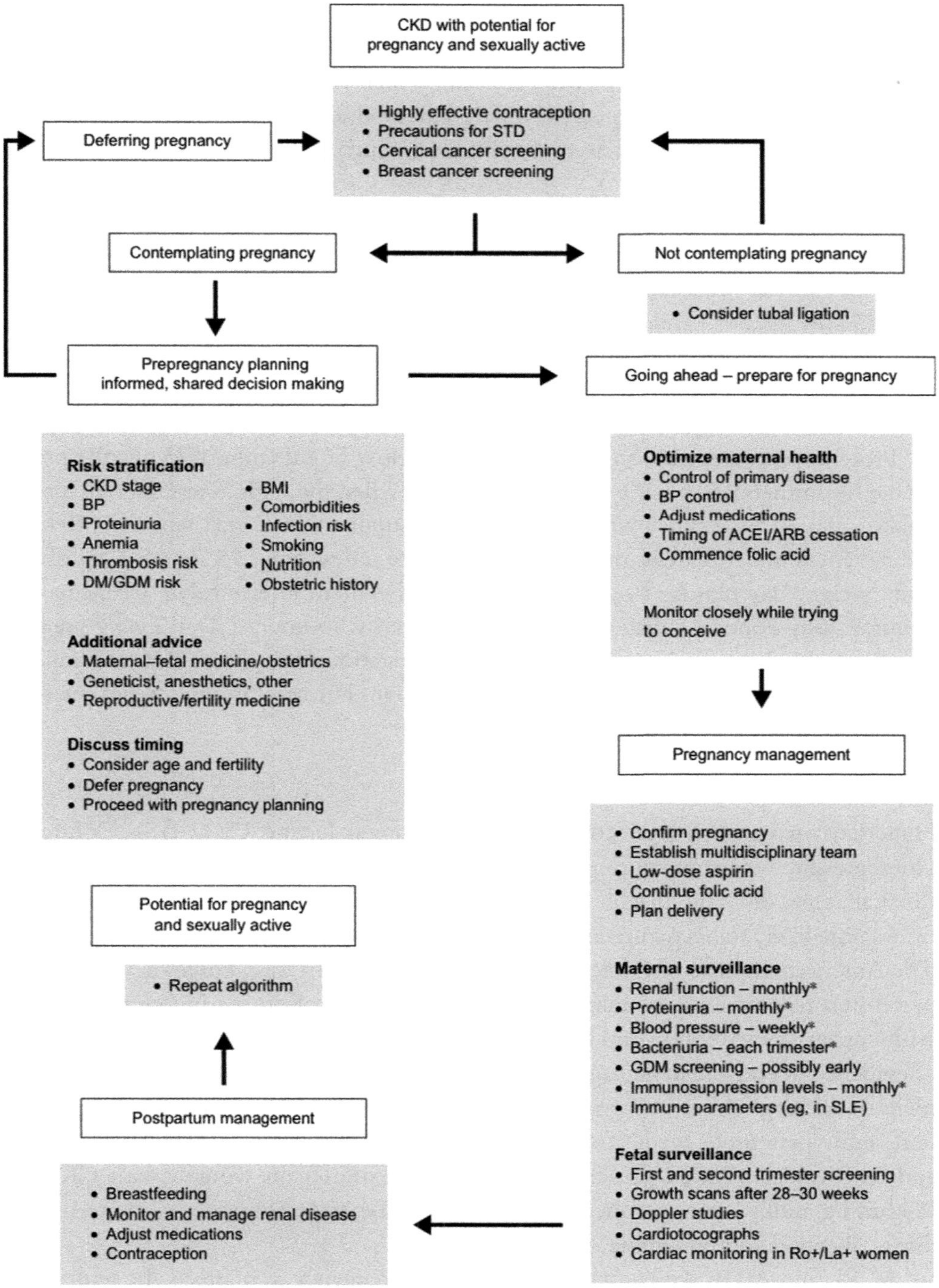

Figure 17.1 Algorithm for pregnancy care in women with CKD [22].

Other conditions associated with kidney disease, like diabetes, need specific attention with women needing to control their glycemic level preconception and during pregnancy for better outcomes [36]. Other conditions such as congenital kidney abnormalities (reflux nephropathy or inherited polycystic kidney disease) need to have regular urine checks, with the treatment of asymptomatic bacteriuria [2,37].

Nutrition

In stages 1−2 CKD, use energy intake recommendation for healthy pregnant women. In stages 3−4, the recommended ended energy intake is 96.2−146.4 kJ/kg/day (23−35 kcal/kg/day), and use the current weight for calculation. If the patient is already on dialysis, American guidelines suggest 125.5−167.4 kJ/kg/day (30−35 kcal/kg/day), while European guidelines suggest 125.5−167.4 kJ/per ideal body weight/day (30−40 kcal/kg/day).

Protein intake depends on GFR. If GFR is below 50 mL/min/1.73 m^2, the protein intake recommendation is 0.6−0.8 g/kg/day and when the GFR is less than 20, protein intake is 0.3−0.5 g/kg/day with ketoacid analog supplementation (1 tablet/8−10 kg). If the patient is under dialysis, different reported case series showed 1.2−1.5 g/kg of ideal body weight/day plus 6−10 g/day showed improvement in electrolyte balance and in another study done in Italian recommended patient who under HD 1.2 g/kg pregestational weight/day and who in PD 1.4 g/kg pregestational weight/day. Regarding lipid intake, we can use the same as nonpregnant women. For macronutrients and micronutrients, see Table 17.3 [39].

Hypertension

Hypertension is a common complication in women having CKD [15]. Controlling blood pressure before and during pregnancy improves outcomes. It is an additive risk factor at all stages of CKD. Piccoli et al. [15] reported that hypertensive women were three times more likely to have a preterm birth versus normotensive women with stage 1 CKD. The American College of Obstetrics and Gynecology 2013 Task Force on Chronic Hypertension recommends avoiding antihypertensive medication use in pregnant women with chronic hypertension and blood pressure less than 160/105 mmHg in the absence of evidence of end-organ damage [23], but for pregnant women with CKD and diabetes mellitus, kidney disease outcomes quality initiative KDOQI guidelines recommend to start antihypertensive medication when blood pressure is more than 140/90 mmHg and to consider a blood pressure target <130/80 mmHg in women with CKD [5]. Regarding antihypertensive medication, the first line and commonly used are methyldopa, labetalol, and nifedipine.

ACEIs and ARBs are contraindicated during pregnancy as it affects the fetus (intrauterine growth restriction, renal dysplasia, oligohydramnios, and fetal death) (Table 17.4) [6,40].

Table 17.3 Nutritional requirements in pregnant CKD women [38].

		Nondialysis (CKD stages 3–5)	HD	PD
Macronutrients				
Calories (Kcal/kg/day)[a]	First	35	30–35	25–30
	Second-third trimester	30–35 (+300 Kcal)	30–35 (+300 Kcal)	25–30 (+300 Kcal)
Proteins (g/kg/day)[a]		0.6–0.8 (+10 g) if uremic syndrome is not controlled, start dialysis	1.2 (+10 g)	1.2 (+10 g)
Micronutrients				
Folic acid (mg/day)		6	2–5	
25-OH vitamin D (IU/day)		1000–2000	1000–2000	
Zinc (mg/day)		15	15	
Iron (mg/day)		20–30	20–30	
Others §				
Electrolytes				
Calcium (mg/day)		<2000	1500–2000	
Phosphate (mg/day)		CKD stages 4–5: 800–1000	1200	
Potassium, mEq/L/day (g)		According to the serum levels	<75 (3 g)	

CKD, chronic kidney disease; *HD*, hemodialysis; *PD*, peritoneal dialysis. Notes: For CKD, patients in stage 1–2 follow the recommendations for healthy pregnancies.

[a]Calculated on prepregnancy weight § including: vitamin C, thiamine, riboflavin, niacin, vitamin B6 and B12. For reference values and a hill list of indications, see Refs. [31] and [47].

Preeclampsia prophylaxis and diagnosis

Preeclampsia, in general, affects 3%–5% of pregnancies, but this rises to 33% if mothers have chronic renal disease [8]. If serum creatinine is more than 2.5 mg/dL, the risk of preeclampsia is over 40% [9]. It is difficult to differentiate between superimposed preeclampsia and exacerbation of chronic disease, especially if mothers have chronic hypertension and proteinuria before pregnancy. The criteria to distinguish are worsening blood pressure (greater than 160/110), worsening proteinuria, other signs of preeclampsia such as hyperuricemia (>5.5 mg/dL), thrombocytopenia (platelet count <100,000 cells/mm^3), elevated lactate dehydrogenase, and elevated liver transaminases [23]. It is recommended to start prophylactic aspirin (50–150 mg/day) starting from 12 weeks of gestation throughout pregnancy especially with advanced CKD regardless of hypertension or proteinuria to decrease the risk of preeclampsia [23,41].

Table 17.4 Summary of antihypertensive medications in pregnancy [25].

	Advantages	Disadvantages
First-line agents		
Nifedipine (PO)	Extended release formulation/ once-daily dosing. Appears to be safe	Edema, headache
Labetalol (PO or IV)	Appears to be safe. Theoretical benefit of α-blockade on placental blood flow	Exacerbation of reactive airway disease
Methyldopa (PO)	Extensive safety data	Fatigue, sedation, poor efficacy, short duration of action
Second-line agents		
Verapamil, diltiazem	Appear to be safe	Limited safety data
Metoprolol (PO or IV)	Extended release formulation/ once-daily dosing	Limited safety data
Hydralazine (PO or IV)	Extensive clinical experience	Maternal tachycardia, delayed maternal hypotension
Nicardipine (IV)	Extensive clinical experience as tocolytic	Requires intravenous administration in a monitored setting
Generally avoided		
Thiazide, loop diuretics	No clear adverse fetal effects	Theoretical concern regarding blunting of physiologic increase in maternal plasma volume
Spironolactone/ aldactone	Effective for primary hyperaldosteronism	Theoretical risk of feminization of male fetuses
Atenolol	None; use other agents	Fetal growth restriction
Nitroprusside (IV)	Rapidly effective for short-term use if other agents fail or are contraindicated	Risk of cyanide poisoning to fetus
Contraindicated		
ACEIs	Risks with early first trimester exposure appear low	Multiple fetal defects with second and third trimester exposure
ARBs	Similar to ACEIs	Similar to ACEIs

PO, by mouth; *IV*, intravenous.

Anemia

Iron deficiency is common during pregnancy, but anemia in case of CKD occurs when eGFR is below 30 mL/min/1.73 m^2 [10]. Erythropoietin is essential in case of chronic anemia to correct iron levels. It is safe during pregnancy and the doses can be increased if hemoglobin is below Hb > 11 g/DL [42].

Venous thromboembolism

If proteinuria is in the nephrotic range or nephrotic syndrome during pregnancy, there is an increased risk of a thrombotic event in addition to pregnancy so consider aspects such as low molecular weight heparin during pregnancy for thromboprophylaxis [12].

Antenatal screening

Standard antenatal screening should be offered to pregnant women with CKD, but should be interpreted cautiously. Screening for these women includes ultrasound nuchal translucency measurement and serum free β-hCG and pregnancy-associated plasma protein A (PAPP-A) in the first trimester [16,43,44]. Serum free β-hCG, unconjugated estradiol, α-fetoprotein, and inhibin A are measured in the second trimester but should be interpreted carefully as their levels can be affected by renal clearance and this might affect test validity. Benachi et al. [45] studied free and total β-hCG in the second trimester to determine those "at risk" of Down syndrome and the result showed that a large number of women with CKD met the criteria for "at risk" using free β-hCG (48% in women with CKD vs. 12% of controls), whereas total hCG gave a nonsignificant difference (25% vs. 15%).

α-fetoprotein, which is mostly used for open neural tube defects, is unaffected by renal disease. PAPP-A, a marker used for SGA babies, is interpretable in early stage CKD but unreliable in the late stages of CKD including dialysis. PAPP-A can also be falsely high in pregnant women using heparin [46,47]. Unconjugated estriol is cleared by dialysis not like α-fetoprotein and β-hCG.

To date, no data are present regarding noninvasive prenatal testing using cell-free DNA in pregnant women with CKD. Future studies are needed to improve interpretation of screening tests in women with CKD.

Fetal monitoring

Feto-maternal medicine should be involved in the follow-up of pregnant women with CKD. Fetal well-being starts from the early detection of any congenital anomalies by doing nuchal translucency ultrasound between 11 and 13 + 6 weeks for gestation, and a detailed anatomy ultrasound between 18 and 24 weeks of gestation. Serial growth scan with amniotic fluid volume and umbilical eatery Doppler is recommended fortnightly from 28 to 30 weeks of gestation in order to detect early fetal growth restriction which can alter the time and mood of delivery [48]. Cardiotocography can be used after 28 weeks of gestation as an additional tool for fetal well-being, especially in women receiving dialysis.

Timing and mode of delivery

There are no studies that recommend an exact time of delivery but in general, rapid worsening of maternal or fetal parameters usually warrants delivery. Vaginal delivery is preferable over Cesarean section if no contraindications [49].

Regarding hypertension in pregnancy, there is a clear recommendation from ACOG 2019: if pregnant women with chronic hypertension are not being treated, delivery is not recommended before 38 weeks of gestation, but if she is on antihypertensive medication and both mother and fetus are stable, delivery is not recommended before 37 weeks of gestation. If the woman develops superimposed preeclampsia without severe features, she can be managed until 37 weeks of gestation, but if she develops severe features, stabilize her and consider delivery at 34 weeks only if both the mother and the fetus are stable. In case of rapid fetal or maternal deterioration at earlier gestation, immediate delivery after ensuring fetal lung maturation is warranted.

Lactation

Breastfeeding is important for infants, and most mothers are encouraged to breastfeed their babies. All antihypertensive medications are safe during lactation and most renal medications are safe, except for mycophenolate or mammalian targets of rapamycin inhibitors due to lack of data. Neonatologists and clinical pharmacists should advise accordingly [50,51].

Contraception

Good advice to patients regarding contraception and CKD is important to optimize disease control and decrease complications. Full counseling regarding each type of contraception including risks and benefits is crucial.

Suitable contraceptive choices for women with CKD include barrier methods, hormonal methods, intrauterine devices, and permanent sterilization if the family is completed. Emergency contraception is also safe for these women.

Progesterone-only contraception including oral, injection, intrauterine devices, or implants are safe and effective for women with CKD. Additional advantage of progesterone-only contraceptives is amenorrhea which may help with the management of iron deficiency anemia for women with CKD.

Combined oral contraception is contraindicated in patients with cardiovascular disease, history of thrombosis, or smoking, and it is relatively contraindicated in women with hypertension, diabetes, or SLE [28].

Conclusion

CKD in pregnancy poses several challenges to women and clinicians, with increasing both maternal and fetal risks. Adverse maternal outcomes include deterioration of renal function, worsening of proteinuria, hypertension, and preeclampsia. On the other hand, fetal risks include fetal loss, fetal growth restriction, and preterm delivery. A multidisciplinary specialized team management is essential to properly care for women with CKD and their babies.

References

[1] Coresh J, Selvin E, Stevens LA, Manzi J, Kusek JW, Eggers P, et al. Prevalence of chronic kidney disease in the United States. JAMA 2007;298(17):2038–47.

[2] Bramham K, Lightstone L. Pre-pregnancy counseling for women with chronic kidney disease. J Nephrol 2012;25(4):450–9.

[3] Wiles KS, Bramham K, Vais A, Harding KR, Chowdhury P, Taylor CJ, et al. Pre-pregnancy counselling for women with chronic kidney disease: a retrospective analysis of nine years' experience. BMC Nephrol 2015;16:28.

[4] Kidney Foundation. NKF KDOQI guidelines: kidney foundation; n.d. Available from: URL: https://kidneyfoundation.cachefly.net/professionals/KDOQI/guidelines_ckd/p4_class_g1.htm.

[5] National Kidney Foundation. KDOQI clinical practice guideline for diabetes and CKD: 2012 update. Am J Kidney Dis 2012;60(5):850–86.

[6] SOMANZ. Guideline for the management of hypertensive disorders of pregnancy. 2014. Available from: URL: https://somanz.org/documents/HTPregnancyGuidelineJuly2014.pdf.

[7] Bharti J, Vatsa R, Singhal S, Roy KK, Kumar S, Perumal V, et al. Pregnancy with chronic kidney disease: maternal and fetal outcome. Eur J Obstet Gynecol Reprod Biol 2016;204:83–7.

[8] Nevis IF, Reitsma A, Dominic A, McDonald S, Thabane L, Akl EA, et al. Pregnancy outcomes in women with chronic kidney disease: a systematic review. Clin J Am Soc Nephrol 2011;6(11):2587–98.

[9] Jones DC, Hayslett JP. Outcome of pregnancy in women with moderate or severe renal insufficiency. N Engl J Med 1996;335(4):226–32.

[10] World Health Organization. Make every mother and child count: the World Health Report. World Health Organization; 2005.

[11] Oz BB, Hackman R, Einarson T, Koren G. Pregnancy outcome after cyclosporine therapy during pregnancy: a meta-analysis. Transplantation 2001;71(8):1051–5.

[12] Basude S, Hein C, Curtis SL, Clark A, Trinder J. Low-molecular-weight heparin or warfarin for anticoagulation in pregnant women with mechanical heart valves: what are the risks? A retrospective observational study. BJOG 2012;119(8):1008–13. discussion 1012-3.

[13] Bili E, Tsolakidis D, Stangou S, Tarlatzis B. Pregnancy management and outcome in women with chronic kidney disease. Hippokratia 2013;17(2):163.

[14] Zhang J-J, Ma X-X, Hao L, Liu L-J, Lv J-C, Zhang H. A systematic review and meta-analysis of outcomes of pregnancy in CKD and CKD outcomes in pregnancy. Clin J Am Soc Nephrol 2015;10(11):1964–78.

[15] Piccoli GB, Cabiddu G, Attini R, Vigotti FN, Maxia S, Lepori N, et al. Risk of adverse pregnancy outcomes in women with CKD. J Am Soc Nephrol 2015;26(8):2011–22.

[16] Imbasciati E, Gregorini G, Cabiddu G, Gammaro L, Ambroso G, Del Giudice A, et al. Pregnancy in CKD stages 3 to 5: fetal and maternal outcomes. Am J Kidney Dis 2007;49(6):753–62.

[17] Jungers P, Chauveau D. Pregnancy in renal disease. Kidney Int 1997;52(4):871–85.

[18] Jungers P, Houillier P, Forget D, Labrunie M, Skhiri H, Giatras I, et al. Influence of pregnancy on the course of primary chronic glomerulonephritis. Lancet 1995;346(8983):1122–4.

[19] Tangren JS, Powe CE, Ankers E, Ecker J, Bramham K, Hladunewich MA, et al. Pregnancy outcomes after clinical recovery from AKI. J Am Soc Nephrol 2017;28(5):1566–74.
[20] Imbasciati E, Pardi G, Capetta P, Ambroso G, Bozzetti P, Pagliari B, et al. Pregnancy in women with chronic renal failure. Am J Nephrol 1986;6(3):193–8.
[21] Shimizu A, Takei T, Moriyama T, Itabashi M, Uchida K, Nitta K. Effect of pregnancy and delivery on the renal function and the prognosis of patients with chronic kidney disease stage 3 caused by immunoglobulin a nephropathy. Intern Med 2015;54(24):3127–32.
[22] Fitzpatrick A, Mohammadi F, Jesudason S. Managing pregnancy in chronic kidney disease: improving outcomes for mother and baby Int J Womens Health 2016;8:273–85.
[23] American College of Obstetricians and Gynecologists. Report of the American college of obstetricians and gynecologists' task force on hypertension in pregnancy. Obstet Gynecol 2013;122(5):1122.
[24] Kuo VS, Koumantakis G, Gallery ED. Proteinuria and its assessment in normal and hypertensive pregnancy. Am J Obstet Gynecol 1992;167(3):723–8.
[25] Kimmel PL, Rosenberg ME, editors. Chronic renal disease. San Diego: Academic Press; 2015.
[26] de Castro I, Easterling TR, Bansal N, Jefferson JA. Nephrotic syndrome in pregnancy poses risks with both maternal and fetal complications. Kidney Int 2017;91(6):1464–72.
[27] Creasy RK, Resnik R. Creasy and Resnik's maternal-fetal medicine| maternal-fetal medicine. Elsevier Saunders; 2014.
[28] ACOG Practice Bulletin No. 203. Chronic hypertension in pregnancy. Obstet Gynecol 2019;133(1): e26–50.
[29] Stratta P, Canavese C, Quaglia M. Pregnancy in patients with kidney disease. J Nephrol 2006;19(2): 135–43.
[30] Davison J, Nelson-Piercy C, Kehoe S, Baker P. Renal disease in pregnancy. Cambridge, U.K: Cambridge University Press; 2014.
[31] National Institute for Health and Care Excellence. Preconception advice and management: clinical knowledge summaries. 2017. Available from: URL: https://www.nice.org.uk/cks-uk-only.
[32] Fredi M, Lazzaroni MG, Tani C, Ramoni V, Gerosa M, Inverardi F, et al. Systemic vasculitis and pregnancy: a multicenter study on maternal and neonatal outcome of 65 prospectively followed pregnancies. Autoimmun Rev 2015;14(8):686–91.
[33] Buyon JP, Kim MY, Guerra MM, Laskin CA, Petri M, Lockshin MD, et al. Predictors of pregnancy outcomes in patients with lupus: a cohort study. Ann Intern Med 2015;163(3):153–63.
[34] Wiles K, Chappell L, Clark K, Elman L, Hall M, Lightstone L, et al. Clinical practice guideline on pregnancy and renal disease. BMC Nephrol 2019;20(1):401.
[35] Webster P, Wardle A, Bramham K, Webster L, Nelson-Piercy C, Lightstone L. Tacrolimus is an effective treatment for lupus nephritis in pregnancy. Lupus 2014;23(11):1192–6.
[36] Bramham K, Rajasingham D. Pregnancy in diabetes and kidney disease. J Ren Care 2012;38(Suppl. 1):78–89.
[37] Williams D, Davison J. Chronic kidney disease in pregnancy. Br Med J 2008;336(7637):211–5.
[38] Esposito P, Garibotto G, Picciotto D, Costigliolo F, Viazzi F, Conti NE. Nutritional challenges in pregnant women with renal diseases: relevance to fetal outcomes. Nutrients 2020;12(3).
[39] Reyes-López MA, Piccoli GB, Leone F, Orozco-Guillén A, Perichart-Perera O. Nutrition care for chronic kidney disease during pregnancy: an updated review. Eur J Clin Nutr 2020;74(7):983–90.
[40] American College of Obstetricians and Gynecologists. Hypertension in pregnancy. Washington, DC: American College of Obstetricians and Gynecologists; 2013. Available from: URL: http://www.spog.org.pe/web/phocadownloadpap/HypertensioninPregnancy.pdf.
[41] Castellano G, Losappio V, Gesualdo L. Update on pregnancy in chronic kidney disease. Kidney Blood Press Res 2011;34(4):253–60.
[42] Hladunewich M, Hercz AE, Keunen J, Chan C, Pierratos A. Pregnancy in end stage renal disease. Semin Dial 2011;24(6):634–9.
[43] ACOG Practice Bulletin No 77. Screening for fetal chromosomal abnormalities. Obstet Gynecol 2007;109(1):217–27.

[44] Australian Department of Health. Clinical practice guidelines - antenatal care. Australian Department of Health; 2013. Available from: URL: http://www.health.gov.au/internet/publications/publishing.nsf/Content/clinical-practice- guidelines-ac-mod1~part-b~chromosomal-abnormalities~background.

[45] Benachi A, Dreux S, Kaddioui-Maalej S, Czerkiewicz I, Fakhouri F, Thervet E, et al. Down syndrome maternal serum screening in patients with renal disease. Am J Obstet Gynecol 2010;203(1):60.e1–4.

[46] Shulman LP, Briggs R, Phillips OP, Friedman SA, Sibai B. Renal hemodialysis and maternal serum triple analyte screening. Fetal Diagn Ther 1998;13(1):26–8.

[47] Valentin M, Muller F, Beaujard MP, Dreux S, Czerkiewicz I, Meyer V, et al. First-trimester combined screening for trisomy 21 in women with renal disease. Prenat Diagn 2015;35(3):244–8.

[48] Turan OM, Turan S, Gungor S, Berg C, Moyano D, Gembruch U, et al. Progression of Doppler abnormalities in intrauterine growth restriction. Ultrasound Obstet Gynecol 2008;32(2):160–7.

[49] Hebert MF, Zheng S, Hays K, Shen DD, Davis CL, Umans JG, et al. Interpreting tacrolimus concentrations during pregnancy and postpartum. Transplantation 2013;95(7):908–15.

[50] Bramham K, Chusney G, Lee J, Lightstone L, Nelson-Piercy C. Breastfeeding and tacrolimus: serial monitoring in breast-fed and bottle-fed infants. Clin J Am Soc Nephrol 2013;8(4):563–7.

[51] Constantinescu S, Pai A, Coscia LA, Davison JM, Moritz MJ, Armenti VT. Breast-feeding after transplantation. Best Pract Res Clin Obstet Gynaecol 2014;28(8):1163–73.

CHAPTER 18

Pregnancy in kidney transplant patients: Considerations and management

Ranajit Chatterjee[1] and Lalit Gupta[2]
[1]Swami Dayanand Hospital, New Delhi, Delhi, India; [2]Anaesthesia and Intensive Care, Maulana Azad Medical College, New Delhi, Delhi, India

Understanding the intersection of kidney transplants and pregnancy

Women with advanced chronic kidney disease (CKD) are often advised to postpone pregnancy until after receiving a kidney transplant due to improved fertility and lower pregnancy complication rates compared to those on dialysis. Despite the benefits of kidney transplantation, pregnancy after transplant remains high-risk, with conditions like preeclampsia occurring in 25%–30% of cases [1]. Factors influencing pregnancy outcomes in transplant recipients include maternal comorbidities, kidney allograft function, the effect of pregnancy on the transplanted kidney, and the impact of maternal health on the fetus, alongside considerations related to immunosuppressive medications and potential teratogenic effects [2]. The number of female kidney transplant recipients in their reproductive years is increasing, underscoring the need for preconception counseling, contraceptive management, and family planning.

The purpose of this chapter is to provide an overview of the considerations and management of pregnancy in kidney transplant patients. The chapter will cover the counseling on contraception, maternal medical management during pregnancy, and the choice of immunosuppressive drugs to prevent fetal toxicity. Additionally, the chapter will discuss the optimal timing of pregnancy, the impact of pregnancy on kidney allograft loss and rejection, and the ideal maternal age of conception.

Managing pregnancy in kidney transplant patients: "Historical consideration"

Addressing pregnancy in kidney transplant patients is crucial as it impacts both the maternal and fetal health. Understanding the complex interplay between maternal health, kidney function, and immunosuppressive medication is vital for optimizing pregnancy outcomes. Healthcare providers need to carefully manage these pregnancies to minimize complications and ensure the well-being of both the mother and the developing fetus.

The Kidney of the Critically Ill Pregnant Woman
ISBN 978-0-443-21473-8
https://doi.org/10.1016/B978-0-443-21473-8.00012-4

The first successful pregnancy in a kidney transplant recipient date back to 1958. Edith Helm, a 23 yearold woman who had received a kidney from her identical twin sister in 1956, achieved this milestone. She delivered a healthy full-term boy weighing 3300 g via caesarean section. Remarkably, Edith's twin sister, Wanda Foster, who had donated the kidney, also went on to have four successful pregnancies after the donation, demonstrating the resilience of both the recipient and the donor [3]. Subsequent successes in pregnancies among kidney transplant recipients have continued to provide encouragement and valuable insights into managing these unique pregnancies.

Preconception counseling and planning

Importance of preconception counseling

Preconception counseling plays a pivotal role in ensuring the well-being of women with kidney transplants who desire pregnancy. The counseling process involves education, risk assessment, medical optimization, and intervention planning before conception [4]. It is essential for several reasons:

1. *Minimizing Poor Pregnancy Outcomes*: The primary goal is to minimize the likelihood of poor pregnancy-related outcomes. This includes reducing risks to both the mother and the fetus.
2. *Education and Informed Decision-making*: It provides individuals and couples with essential information about the potential challenges and risks associated with pregnancy after kidney transplantation. Informed decision making is crucial.
3. *Medication Management*: Preconception counseling allows healthcare providers to evaluate the patient's medication regimen and make necessary adjustments to ensure safety during pregnancy.
4. *Timing and Planning*: It helps in determining the right timing for pregnancy, considering factors such as graft stability and the patient's overall health.

Assessing the patient's readiness for pregnancy

Assessing readiness for pregnancy involves evaluating various aspects of the patient's health and circumstances. This assessment includes [5]:

1. Overall Health: Ensuring that the patient's overall health is stable and that any underlying conditions, besides kidney transplantation, are well managed.
2. Psychosocial Factors: Assessing the patient's emotional and psychological readiness for the challenges of pregnancy and potential complications.
3. The stability of the transplanted kidney is a critical factor. Physicians will monitor kidney function, looking for signs of rejection or other issues that may affect pregnancy.

Reviewing the patient's immunosuppressive medication regimen

Immunosuppression management is a critical consideration before conception in kidney transplant recipients. These medications can cross the placenta, posing potential risks to the fetus. Most immunosuppressive drugs used during pregnancy fall into categories C and D according to the Food and Drug Administration's pregnancy safety classification (Table 18.1) [6–10]. Short acting corticosteroids are most commonly used in kidney transplant regimens, but they can also lead to adverse effects for both the fetus and mothers, including developmental delays and diabetes.

Careful management of immunosuppressive medications and their potential risks is crucial for kidney transplant recipients considering pregnancy. Reviewing immunosuppressive medications is essential to minimize risks to both the mother and the fetus. Consultation with healthcare providers and adherence to safety programs are essential for ensuring both maternal and fetal well-being. Medication adjustments may be necessary to balance the need for immunosuppression and the safety of the pregnancy.

Discussing potential risks and complications

Open and honest discussions about potential risks and complications are crucial. This includes understanding the impact of immunosuppressive medications on the fetus, the risk to the kidney allograft, and other maternal complications associated with pregnancy after kidney transplantation [11].

Preconception counseling empowers patients to make informed decisions about family planning and ensures that pregnancies occur under the best possible circumstances, enhancing the chances of a successful pregnancy while safeguarding the health of both the mother and the transplanted kidney.

Maternal outcomes

Hypertension and preeclampsia: Kidney transplant recipients who become pregnant face a relatively high risk of hypertension, with approximately 54.2% experiencing this condition [12]. Additionally, maternal age plays a role in increasing the incidence of hypertension and preeclampsia in this population. These women are also at a greater risk of developing preeclampsia and gestational diabetes compared to the general population with preeclampsia rates ranging between 21.5% and 31% [13].

Distinguishing between new onset hypertension and preeclampsia can be challenging, especially when proteinuria results from various sources like acute allograft rejection or infection. Treatment for hypertension during pregnancy in kidney transplant recipients largely mirrors that in the general population, with beta-blockers and calcium channel blockers being more effective than methyldopa in reducing the risk of severe hypertension [14].

Table 18.1 Summarizes information on the use of immunosuppressive medications during pregnancy.

Medication	FDA pregnancy category	Risks and effects on fetus and mother	Precautions and considerations
Short acting corticosteroids	Not specified	In utero exposure to high doses associated with fetal issues like cleft palate, developmental delay, and thymic hypoplasia. Adverse effects in mothers include premature rupture of membranes, diabetes, and hypertension.	Monitor and manage doses carefully during pregnancy.
Mycophenolate mofetil (MMF)	Category D	Teratogenic risk, associated with congenital malformations (e.g., cleft lip, heart defects), increased risk of miscarriage, preterm birth, and small for gestational age infants. FDA has a risk evaluation and mitigation strategies program.	Switch to alternative medications at least 6 weeks before conception. Emphasize contraception and conception planning while on MMF.
Azathioprine	Category D	Limited placental transfer, low risk of intrauterine exposure.	Consider switching to azathioprine before conception.
Calcineurin inhibitors	Not specified	Metabolism increases during pregnancy, may require a 25%–50% dose increase.	Adjust dosages as needed, especially in cases of anemia and hypoalbuminemia.
Sirolimus	Category C	Limited data on congenital malformations, few successful pregnancies reported.	Limited information available; exercise caution.
Immunoglobulin	Category C	Cross placenta only after 32 weeks and not proven embryotoxic.	Safe to use during and after pregnancy.
Other immunosuppressive agents	Not specified	Limited information on use during pregnancy, some promising reports (e.g., Belatacept, Basiliximab).	Case by case assessment, consult healthcare provider.

Category C implies adverse effects in animal studies or insufficient data in humans, while category D suggests evidence of fetal risk in humans.

Medications like labetalol, nifedipine, and amlodipine are commonly used and considered safe during pregnancy. However, atenolol usage during the first trimester has been associated with low birth weight [15]. Angiotensin-converting enzyme inhibitors and angiotensin receptor blockers are contraindicated due to the higher risk of fetal kidney injury and should ideally be discontinued at least 6 weeks before conception [16].

Kidney allograft rejection

Pregnancy is not associated with a higher risk of rejection or allograft loss in kidney transplant recipients with stable kidney function. However, pregnant recipients with an unstable kidney function face risks of acute graft rejection (1.3%9.4%) and graft loss (5.6%9.2%), particularly if pregnancy occurs within the first and second year after transplantation. Rates of pregnancy-related allograft loss vary among studies, possibly influenced by factors like hypertension, proteinuria, and acute kidney injury (AKI) [17].

Moderate to severe allograft dysfunction, defined by creatinine levels exceeding 1.3 and 1.9 mg/dL, respectively, and proteinuria exceeding 500 mg/dL, increase the risk of graft failure associated with pregnancy. While there is some immunoregulation associated with pregnancy, acute rejection remains possible. In cases of suspected acute rejection, kidney biopsy with ultrasound guidance, preferably during the first and second trimesters, can confirm the diagnosis [17,18].

If rejection during pregnancy is confirmed, high dose corticosteroids and intravenous immunoglobulin can be used; however, data on the use of antithymocyte globulin, basiliximab, and rituximab during pregnancy are limited. Predictors of worsening allograft function include a higher estimated GFR before and after pregnancy, as well as acute rejection within 3 months of the postpartum period [19].

Pregnancy-related acute kidney injury in kidney transplant patients

Pregnancy-related AKI in kidney transplant recipients is influenced by several risk factors. These factors include having a solitary kidney, susceptibility to volume depletion due to impaired autoregulation, immunological risk factors, abdominal compartment syndrome risk based on transplant history, certain medications' nephrotoxic effects (such as calcineurin inhibitors), increased urinary reflux during pregnancy, urinary retention, exposure to nephrotoxic antimicrobials and immunoglobulins, vulnerability to viral infections (e.g., polyomavirus nephropathy and cytomegalovirus [CMV]), potential recurrence of conditions like glomerulonephritis and thrombotic disorders, hematological causes (e.g., antiphospholipid syndrome and hypercoagulable states), vascular issues (e.g., renal artery stenosis), and an elevated risk of posttransplant lymphoproliferative disorder. These factors collectively contribute to the increased complexity and potential challenges in managing kidney transplant recipients during pregnancy [20,21].

Table 18.2 The diagnostic criteria and considerations for pregnancy-related acute kidney injury (AKI) in kidney transplant recipients.

Diagnostic criteria for pregnancy-related AKI	Key points and considerations
Traditional criteria	An acute rise in creatinine by >0.5 mg/dL from baseline within 48 h has been proposed as a defining criterion for pregnancy-related AKI.
Existing AKI criteria (RIFLE, AKIN, KDIGO)	These widely used criteria in the general population (based on creatinine, eGFR, and urine output changes) have not been validated for diagnosing pregnancy-related AKI in kidney transplant recipients.
Serum creatinine	Serum creatinine remains the most common and accessible biomarker for diagnosing AKI in kidney transplant recipients during pregnancy.
Other biomarkers (e.g., cystatin C, NGAL, KIM1)	Biomarkers like cystatin C, NGAL, and KIM1 are not routinely used in clinical practice for AKI diagnosis in this population and lack validation.
Hematuria and proteinuria	Hematuria and proteinuria can be used as additional indicators of AKI, especially when kidney function is not regularly assessed.
Regular assessment	Routine assessment of proteinuria and hematuria should be part of the standard care for pregnant kidney transplant recipients to detect early signs of AKI.
Etiologies of AKI	The causes of AKI can vary during different stages of pregnancy, and it's essential to consider these factors when evaluating AKI in this population.

Due to the lack of validated criteria for pregnancy-related AKI in kidney transplant recipients, a comprehensive assessment combining various clinical indicators, including serum creatinine, hematuria, and proteinuria, is crucial for timely detection and management [22–25] (Table 18.2).

Management of pregnancy-related AKI in kidney transplant recipients

The management of pregnancy-related AKI in kidney transplant recipients is a complex and collaborative effort involving various medical specialties and tailored approaches based on the individual's condition and underlying causes [2,26,27]. (Table 18.3).

Dialysis in pregnancy is often increased for individuals who were already on dialysis before getting pregnant. It is recommended to have more than 36 h of dialysis per week to improve outcomes for the baby. However, the decision to start dialysis during pregnancy is usually based on concerns about the harmful effects of urea on the fetus, rather than the usual reasons for dialysis.

Table 18.3 Management of pregnancy-related AKI in kidney transplant recipients.

Management of pregnancy related AKI in kidney transplant recipients	Key points and considerations
Multidisciplinary team approach	Involves transplant nephrologists, maternal fetal medicine specialists, transplant surgeons, and neonatologists.
Blood pressure monitoring	Regular monitoring, especially in the second and third trimester. Hypertension with or without proteinuria requires close attention and may necessitate hospital admission and fetal Doppler monitoring.
Identifying and addressing underlying causes	Prompt identification and reversal of the cause of kidney injury. Specific measures based on the cause (e.g., intravenous fluids for prerenal AKI, blood transfusion for anemia-related ischemic AKI).
Electrolyte monitoring and correction	Close monitoring and correction of serum electrolyte imbalances.
Diuretics	Caution in using diuretics due to potential risks of volume depletion or preeclampsia. Discussion among the care team is essential.
Nutritional support	Providing nutritional support is crucial for pregnant kidney transplant recipients with AKI.
Urinary tract infections	Prompt treatment of urinary tract infections and pyuria with appropriate antibiotics, guided by susceptibility studies.
Herpes infection	Caesarean section should be considered if the recipient has primary herpes infection to prevent fetal transmission.
Immunosuppression adjustments	Reduction of immunosuppression levels and antiviral treatment for infections. Consideration of plasmapheresis, eculizumab, and rituximab for conditions like thrombotic thrombocytopenic purpura and atypical hemolytic uremic syndrome.
Glomerulonephritis	Treatment with steroids and immunosuppressive therapy for glomerulonephritis, with a careful assessment of risks and benefits during pregnancy.
Immunosuppressive medications	Azathioprine and calcineurin inhibitors (cyclosporine and tacrolimus) are commonly used and not teratogenic. Frequent monitoring of drug trough levels is essential during pregnancy.
Rejection treatment	Management depends on biopsy findings (antibody mediated or acute cellular rejection), donor specific antibodies, and infection history. High-dose steroids as first-line treatment. Consider escalating immunosuppression or monoclonal antibodies after thorough patient discussions.
Kidney replacement therapy	Initiate when indicated if kidney function continues to deteriorate despite efforts to reverse AKI with consideration for prompt and safe delivery of the fetus.

The optimal urea threshold for starting dialysis in pregnancy is still unknown. If serum urea concentrations are above 15 mmol/L, discussions are recommended to determine the appropriate timing and concentration for starting dialysis. For those who are new to dialysis during pregnancy, it is important to consider residual kidney function and start with gentle dialysis sessions, gradually increasing based on individual parameters. There is no additional benefit shown for long, frequent dialysis in this group [27].

Pregnancy outcomes

The mean gestational age for kidney transplant recipients is approximately 34.9 weeks, compared to 38.7 weeks in the general population. Most pregnancies in kidney transplant recipients result in live births, with a rate between 71% and 73%. However, it's worth noting that this finding may be influenced by a selection bias, as a healthier transplant population may be more inclined to pursue pregnancy compared to the general population [12].

Some studies have reported successful pregnancies and live births in kidney transplant recipients who used in vitro fertilization, but it's unclear if this significantly impacts the increasing rates of live births in this population. Overall, while kidney transplant recipients may face higher risks and challenges during pregnancy, the majority still experience successful pregnancies and live births. Not only the risk of ectopic pregnancy but also the incidence of caesarean section may be increased than general population in transplanted women [28].

Fetal outcomes

Preterm Delivery: Preterm delivery (before 37 weeks of gestation) occurs in a significant proportion of pregnancies among kidney transplant recipients, ranging from 43.1% to 65.4%. This highlights the increased risk of premature birth in this population [12].

Risk of Miscarriages and Stillbirths: Interestingly, kidney transplant recipients have a lower reported risk of miscarriages compared to the general population. However, the incidence of stillbirths in this group varies from 2.5% to 5.1%, which is higher than in the general population [12].

Birth Weight and Neonatal Mortality: The mean birth weight of infants born to transplant recipients tends to be lower than that of the general population [29]. Additionally, neonatal mortality rates are higher between years 2 and 3 after kidney transplantation. These findings indicate the need for careful monitoring and support for neonates born to kidney transplant recipients.

Risk of infections

Pregnant renal transplant recipients are at a higher risk of infections, particularly urinary tract infections (UTIs) and acute pyelonephritis, due to the use of immunosuppressive medications. UTIs are common in these women, and screening should be done regularly. Asymptomatic bacteria should be treated with antibiotics, and prophylaxis should be continued throughout pregnancy. Common options include oral beta lactams (with consideration of resistance among gram-negative pathogens) or nitrofurantoin (with avoidance during the first trimester and at term if other options are available). Monitoring kidney function and urine culture after treatment is advised to confirm the elimination of bacteriuria [30,31].

CMV infection can be transmitted to the fetus, leading to various complications. Herpes simplex infection in the mother can increase the risk of abortion and transmission to the child during birth. Hepatitis B and C can also be transmitted to the new born, but preventive measures such as immunoglobulin and vaccines can significantly reduce the risk [5].

Immunization

Immunizations for pregnant kidney transplant recipients are similar to those for pregnant women in general. Some vaccines are recommended before pregnancy, while others can be given during pregnancy. Studies have shown that vaccines are safe for kidney transplant recipients. Tetanus and pertussis vaccines are recommended during the third trimester. Influenza vaccination is recommended to protect both the mother and the infant. Live vaccines are not recommended for kidney transplant recipients [31]. The use of the MMR vaccine in this population is not well studied. Kidney transplant candidates should receive the MMR vaccine before transplantation. If a kidney transplant recipient is nonimmune and exposed to measles, they should receive intravenous immunoglobulin prophylaxis [32].

Labor and delivery

It's important to consider the potential risks and benefits associated with each method of delivery in transplant patients. For vaginal delivery, it is generally recommended for kidney transplant recipients. However, caution should be exercised in individuals with kidney impairment due to the potential for fluid retention. Induction of spontaneous labor is safe with gestational age $\geq$38 weeks, if no other obstetrical complications exist. Oxytocin can be used for induction, but close monitoring is necessary [5].

On the other hand, caesarean delivery rates tend to be higher in kidney transplant recipients, often due to pregnancy complications such as preeclampsia or fetal growth restriction. During a caesarean section, there is a risk of trauma to the transplanted ureter and kidney allograft, estimated to be between 1% and 2%. Therefore, careful consideration of intraoperative risks to the allograft is essential. To minimize the risk to the allograft, a midline skin incision is recommended to avoid the allograft's location.

Additionally, an abdominal ultrasound is advised to verify the kidney allograft's integrity. Recipients on steroid-based immunosuppression may require stress dose steroids during the procedure. Stress dose steroids are recommended for recipients on steroid-based immunosuppression [33].

Postpartum care and monitoring

Postpartum care for kidney transplant recipients is critical to ensure the well-being of both the mother and the new born. Close monitoring of kidney function in the postpartum period is essential. Kidney transplant recipients may experience fluctuations in kidney function, and any abnormalities should be promptly addressed. Resuming or adjusting immunosuppressive therapy is a key consideration in the postpartum period [34]. The challenge is to find the right balance between maintaining adequate immunosuppression to prevent organ rejection and minimizing potential risks to the new born through breastfeeding. The decision should involve close collaboration between the transplant team and healthcare providers specializing in maternofetal medicine.

Lactation

Lactation is an option for many transplant recipients, with considerations for immunosuppression medications and their potential effects on breastfeeding infants. Certain immunosuppressive medications may be compatible with breastfeeding, while others may carry higher risks. Healthcare providers should guide the patient on the best approach for their specific situation. Low levels of immunosuppression drugs have been detected in breast milk, and the safety of breastfeeding while taking maintenance immunosuppression drugs such as prednisone, azathioprine, and calcineurin inhibitors has been reported. In general, the benefits of breastfeeding often outweigh the potential exposure risks to infants [35].

Conclusion

In the evolving landscape of pregnancies among kidney transplant recipients, a remarkable progress has been achieved through the collective efforts of multidisciplinary teams and advancements in medical science over the past 5 decades. It is imperative to acknowledge that women with kidney transplants encounter elevated risks for both maternal and fetal complications despite the instances of successful pregnancies in solid organ transplant recipients. The susceptibility of kidney transplant recipients to AKI during pregnancy underscores the significance of recognizing the underlying risk factors predisposing them to prerenal, intrinsic, and postrenal insults. Timely diagnosis and adept management are paramount. This multidisciplinary team is essential to ensure favorable outcomes for both mother and child. As we continue to expand our understanding of this complex intersection of transplantation and pregnancy, ongoing research and collaboration will pave the way for improved care, better outcomes, and brighter prospects for women embarking on the remarkable journey of motherhood after kidney transplantation.

References

[1] Hendren EM, Reynolds ML. Pregnancy in renal transplant recipients: histopathology provides new insight. Kidney Int Rep 2022;7(1):6–8. Available from: https://doi.org/10.1016/j.ekir.2021.11.007.

[2] Gonzalez Suarez ML, Parker AS, Cheungpasitporn W. Pregnancy in kidney transplant recipients. Adv Chronic Kidney Dis November 2020;27(6):486498. https://doi.org/10.1053/j.ackd.2020.06.004. PMID: 33328065.

[3] Murray JE, Reid DE, Harrison JH, Merrill JP. Successful pregnancies after human renal transplantation. N Engl J Med 1963;269:341–3. https://doi.org/10.1056/nejm196308152690704.

[4] Ponticelli C, Zaina B, Moroni G. Planned pregnancy in kidney transplantation. A calculated risk. J Pers Med September 26, 2021;11(10):956. https://doi.org/10.3390/jpm11100956.

[5] Shah S, Verma P. Overview of pregnancy in renal transplant patients. Int J Nephrol 2016;2016: 4539342. https://doi.org/10.1155/2016/4539342.

[6] Chandra A, Midtvedt K, Asberg A, Eide IA. Immunosuppression and reproductive health after kidney transplantation. Transplantation 2019;103:e325e333.

[7] Lessan Pezeshki M. Pregnancy after renal transplantation: points to consider. Nephrol Dial Transpl 2002;17:703707.

[8] Sifontis NM, Coscia LACS, Lavelanet AF, Moritz MJ, Armenti VT. Pregnancy outcomes in solid organ transplant recipients with exposure to mycophenolate mofetil or sirolimus. Transplantation 2006; 82:16981702.

[9] Hou S. Pregnancy in renal transplant recipients. Adv Chronic Kidney Dis 2013;20:253259.

[10] Olyaei AJ, Thi K, deMattos AM, Bennett WM. Use of basiliximab and daclizumab in kidney transplantation. Prog Transpl 2001;11(quiz 3839):3337.

[11] Farahi N D, Zolotor A. Recommendations for preconception counseling and care. Am Fam Physician 2013;88(8):499–506.

[12] Deshpande NA, James NT, Kucirka LM, et al. Pregnancy outcomes in kidney transplant recipients: a systematic review and meta analysis. Am J Transpl 2011;11:23882404.

[13] Mortiz M, Constantinescu S, Coscia LA, Kliniewski D. Transplant pregnancy registry international 2017 annual report. Philadelphia, PA. 2017. p. 1–21.

[14] Abalos E, Duley L, Steyn DW, Gialdini C. Antihypertensive drug therapy for mild to moderate hypertension during pregnancy. Cochrane Database Syst Rev 2018;10:CD002252. https://doi.org/10.1002/14651858.

[15] Mito A, Murashima A, Wada Y, et al. Safety of amlodipine in early pregnancy. J Am Heart Assoc 2019;8:e012093. https://doi.org/10.1161/JAHA.119.012093.

[16] Bullo M, Tschumi S, Bucher BS, Bianchetti MG, Simonetti GD. Pregnancy outcome following exposure to angiotensin converting enzyme inhibitors or angiotensin receptor antagonists: a systematic review. Hypertension 2012;60:444450.

[17] Gorgulu N, Yelken B, Caliskan Y, Turkmen A, Sever MS. Does pregnancy increase graft loss in female renal allograft recipients? Clin Exp Nephrol 2010;14:244247.

[18] Majak GB, Reisaeter AV, WeedonFekjaer H, Henriksen T, Michelsen TM. The effect of pregnancy on the long term risk of graft loss, cardiovascular disease, and death in kidney transplanted women in Norway: a retrospective cohort study. Transplantation 2018;102:e391e396.

[19] Kent R, Shapiro R. Pregnancy and kidney transplantation. In: Sachdeva M, Miller I, editors. Obstetric and gynecologic nephrology. Cham: Springer; 2020. p. 193202.

[20] Taber-Hight E, Shah S. Acute kidney injury in pregnancy. Adv Chronic Kidney Dis 2020;27(6): 455–60.

[21] Vinturache A, Popoola J, Watt-Coote I. The changing landscape of acute kidney injury in pregnancy from an obstetrics perspective. J Clin Med 2019;8(9).

[22] Bellomo R, Ronco C, Kellum JA, Mehta RL, Palevsky P. Acute renal failure-definition, outcome measures, animal models, fluid therapy and information technology needs: the Second International Consensus Conference of the Acute Dialysis Quality Initiative (ADQI) Group. Crit Care 2004; 8(4):R204−12.

[23] Mehta RL, Kellum JA, Shah SV, et al. Acute kidney injury network: report of an initiative to improve outcomes in acute kidney injury. Crit Care 2007;11(2):R31.

[24] Vijayan M, Avendano M, Chinchilla KA, Jim B. Acute kidney injury in pregnancy. Curr Opin Crit Care 2019;25(6):580−90.

[25] de Buhren H, Hertig A. Staging pregnancy-related acute kidney injury according to kidney disease: improving global outcomes guidelines: what are the barriers? Nephrol Dial Transplant 2021;36(6): 959−61.

[26] Wiles K, Chappell L, Clark K, et al. Clinical practice guideline on pregnancy and renal disease. BMC Nephrol 2019;20:401. https://doi.org/10.1186/s12882-019-1560-2.

[27] Hladunewich M, Hou S, Odutayo A, et al. Intensive hemodialysis associates with improved pregnancy outcomes: a Canadian and United States cohort comparison. J Am Soc Nephrol 2014;25:1103e9.

[28] Osterman MJ, Martin JA. Trends in lowrisk cesarean delivery in the United States, 1990 through 2013. Natl Vital Stat Rep 2014;63:116.

[29] Sanghavi M, Rutherford JD. Cardiovascular physiology of pregnancy. Circulation 2014;130: 10031008.

[30] Moore C. Transplant and pregnancy. In: Parajuli S, Aziz F, editors. Kidney transplant management. Cham: Springer; 2019. p. 179191.

[31] Liang JL, Tiwari T, Moro P, et al. Prevention of pertussis, tetanus, and diphtheria with vaccines in the United States: recommendations of the advisory committee on immunization practices (ACIP). MMWR Recomm Rep (Morb Mortal Wkly Rep) 2018;67:144.

[32] Duchini A, Goss JA, Karpen S, Pockros PJ. Vaccinations for adult solid organ transplant recipients: current recommendations and protocols. Clin Microbiol Rev 2003;16:357364.

[33] Wiles K, Tillet A, Harding K. Solid organ transplantation in pregnancy. Obstet Gynaecol 2016;18: 189197.

[34] Yadav A, Salas MAP, Coscia L, Basu A, Rossi AP, Sawinski D, Shah S. Acute kidney injury during pregnancy in kidney transplant recipients. Clin Transplant May 2022;36(5):e14668. https:// doi.org/10.1111/ctr.14668.

[35] Constantinescu S, Pai A, Coscia LA, Davison JM, Moritz MJ, Armenti VT. Breastfeeding after transplantation. Best Pract Res Clin Obstet Gynaecol 2014;28:11631173.

Index

Note: "Page numbers followed by *f* indicate figures, *t* indicate tables, and *b* indicate boxes."

D

E

R

S

T

U

V

W

9780443214738